Judith Bunting

GENETICS

BOXTREE

First published in the UK 1994
by BOXTREE LIMITED, Broadwall House,
21 Broadwall, London SE1 9PL

1 3 5 7 9 10 8 6 4 2

Edited by Miranda Smith
Design by Julian Holland Publishing
Artwork by Steve Seymour
Picture research by Dee Robinson
Cover design by Pinpoint Design Company

1-85283-337-8

A catalogue record for this book is available from the British Library

Acknowledgements
Boxtree would like to thank Dana Purvis, the editor of Tomorrow's World for her advice and assistance and MRC Human Genome Mapping Project Resource Centre for information supplied. The photographs that appear in the book were obtained from the following sources:
Pp 4-5 ZEFA Picture Library, p 6 (top left and right) Hulton Deutsch, (bottom left) Camera Press, p 7 Janine Wiedel, p 10 Science Photo Library, p 11 (top and bottom) Science Photo Library, p 12 (top) ZEFA Picture Library, (bottom) Biophoto Associates, p 14 (top) Biophoto Associates, (bottom) Science Photo Library, p 15 (top) Frank Spooner Pictures, (bottom) Science Photo Library, p 16 Promega Corporation, p 17 (top) Monsanto, (middle) R.W. Gibson, (bottom) J. Holland, p 18 (top) Science Photo Library, (bottom) Pharmaceutical Proteins, p 19 (top right) Wella (J. Holland), (bottom left) Holt Studios, (bottom right) Yves Poirier, Michigan State University, p 20 (bottom right) Holt Studios, p 21 (top) I Wyman/Sygma, (bottom) Animal Aid, p 22 (top) David Dickel, (bottom) Dr Cano, California Polytechnic, p23 (top) United International Pictures, (middle) Quagga Project/South Africa Museum, (bottom) C.D. Last from an original drawing by Werner, p 24 Science Photo Library (top and bottom), p 25 (top) The Bridgeman Art Library IND54453, (bottom) Royal Geographical Society, p 26 (bottom) Biophoto Associates, p 27 (top and bottom) Science Photo Library, (middle left) Professor A J Jeffreys/ Department of Genetics/University of Leicester, (middle right) Associated Press, p 28 (top) Brian E Rybolt/Impact, (bottom left and right) Science Photo Library, p 29 Mike Etherington, Central Middlesex Hospital, p 30 (top left) Image Bank, p 31 (top left) Art Directors, (top and bottom right) Tony Stone, (bottom left) Panos Pictures, p 32 ZEFA Picture Library, p 33 Science Photo Library, pp 34-35 Steve Uzzell/College of Physicians & Surgeons, Columbia University, p 38 (top) Science Photo Library, (bottom) Dr Modell p 39 (top) John Watney, (bottom) Science Photo Library, p 40 (top) Frank Spooner Pictures, (bottom) Science Photo Library, p 42 (top and bottom) Science Photo Library, (middle) ZEFA, p 43 (top) Monsanto, (bottom) Science Photo Library, p 44 (left) Camera Press, (right) Science Photo Library, p 45 National Medical Slide Bank.

CONTENTS

INTRODUCTION

IN THE EARLY 1950s, two keen young scientists, Francis Crick and James Watson, cracked the biggest problem in human biology – they deduced the shape and structure of our genes. Research has built on that discovery so swiftly that already, well within Watson and Crick's own lifetimes, genes are being manipulated in ways that have astonishing implications for all our lives.

Genes are the chemicals that control who and what we are. Our education and environment influence how we develop, but they work on the bedrock of our genetic material.

At the same time as genes define our differences, they govern the universal basics of life. As much as 99 per cent of our genetic material is exactly the same as the chimpanzee's; just a 1 per cent difference accounts for language, culture, art and science. As yet, we have no idea how most of our genes work, but a huge amount of international scientific effort is being put into reading the human genome – the entire blueprint for a human being. With a little luck and a lot of patience, this project will succeed in finding some answers.

Where genetic diseases are concerned, answers are coming thick and fast. Barely a month passes without an announcement that the gene responsible for yet another inherited disorder has been discovered. Gradually, these discoveries are being put to work diagnosing inherited conditions and designing treatments (like gene therapy) that promise to challenge some of the major causes of illness and early death which we face in the 20th century.

The basic similarity of genetic material in all living things makes it possible to move genes around. One early experiment transferred a gene from a firefly to a plant and made it glow in the dark. Now fish genes work in tomatoes, genes from bacteria make plastic in potatoes, and human genes are used to make drugs in sheep.

The ability to manipulate the inheritance of the human race is an awesome responsibility, and the very power of genetics demands that it is carefully controlled. Gene therapy will save lives, but should be used with caution. Even perfecting the performance of agricultural crops could have unexpected and unwanted knock-on effects.

The speed of research since the early 1950s has whisked us to the brink of a new scientific revolution. Genetics will be the most important science of the 21st century, and its helical strands are likely to reach out into all our lives.

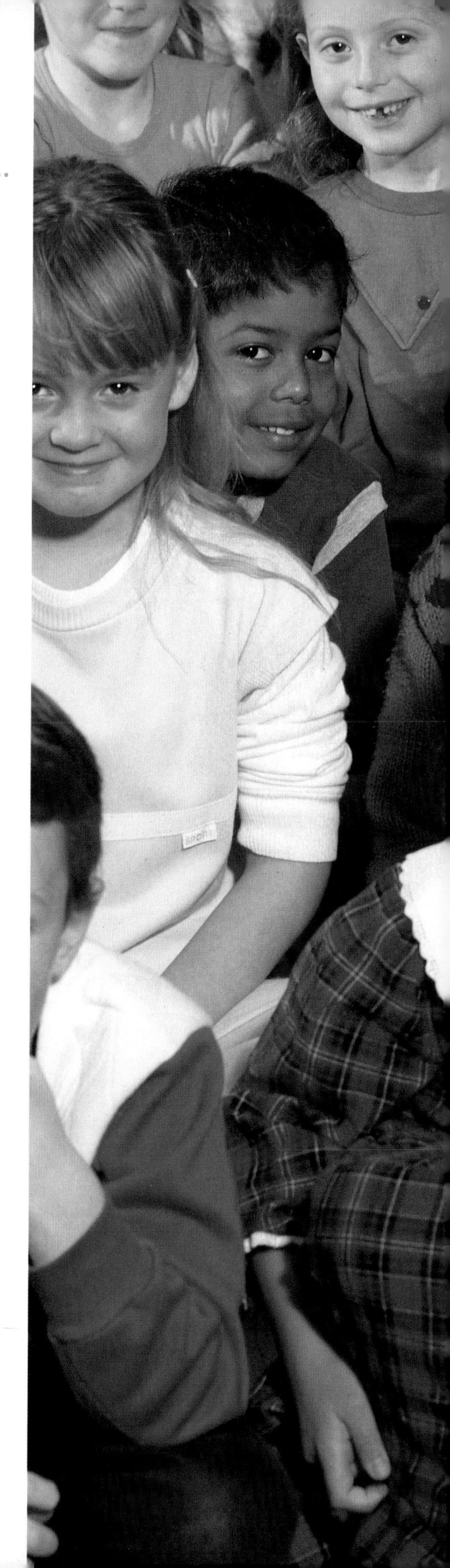

▷ *Within the lifetime of these children, geneticists will read every letter of the human genetic code. What they make of it will affect our lives in more ways than we can currently imagine.*

RUNNING IN THE FAMILY

MOST FAMILIES have some known characteristics that are passed on from generation to generation. It might be a bad temper or a big nose, but everyone in the family knows where it came from.

△ *The three Kennedy brothers, Edward, John and Robert (above) were alike in many ways. Look at their eyes, their noses and their even, white teeth. These characteristics were clearly inherited from their father, Joseph (above left). Robert's son Joe (left) may have curly hair, but he has definitely inherited the Kennedy smile.*

Regular patterns of inheritance were first noticed in the 19th century by Gregor Mendel, a meticulous monk from Moravia with a keen interest in the monastery garden. By cross-breeding thousands of pea plants and following the way that characteristics like the colour of pea flowers were inherited, Mendel discovered that different inherited characteristics follow different patterns as they pass from generation to generation.

The patterns that Mendel observed depend on what became known as the plant's genes. Flower colour in a pea plant is determined by just two genes; let us call the gene for purple flowers P and the gene for white flowers W.

When a pure breeding plant with purple flowers (and the colour genes P-P) is crossed with a pure breeding plant with white flowers (and the colour genes W-W), the next generation all inherit one colour gene from each parent plant. But, despite the fact they all carry one 'white' gene, they all have purple flowers.

If these second generation plants are bred within the group to produce a third generation, three-quarters of the new plants will have at least one purple gene (those with genes P-P, P-W, W-P), and they will have purple flowers. The rest, with two white genes, have white flowers (W-W).

On this evidence the gene for purple flowers is said to be a dominant gene because a plant needs only one copy of it to end up with purple flowers. The gene for white flowers is recessive. That means that for a plant to have white flowers, it must have two copies of the 'white' gene.

Very often, though, genes are not completely dominant or recessive. For instance, red campions crossed with white campions produce pink campions because neither red-flower genes nor white-flower genes dominate, and the two flower colour genes work together.

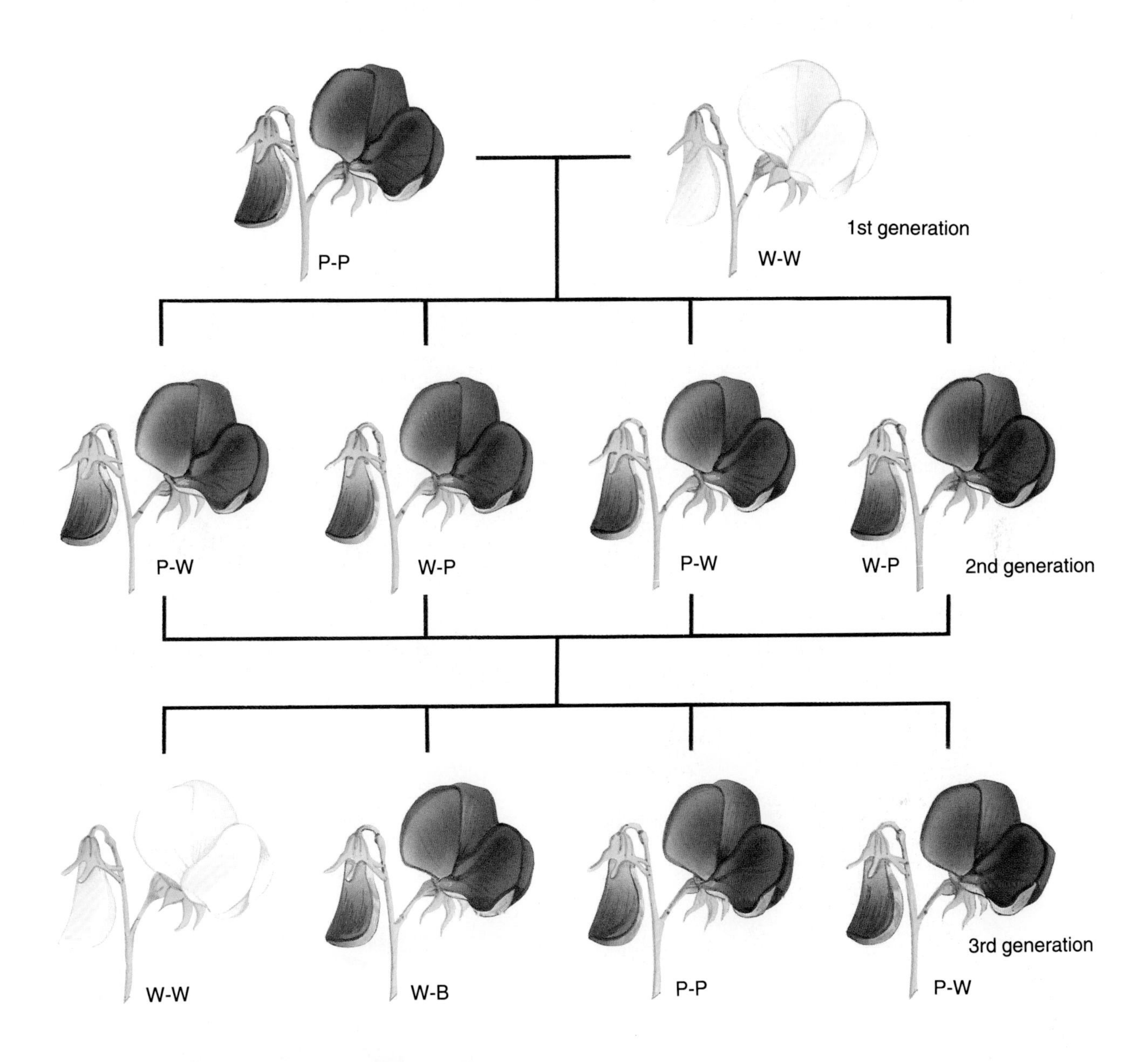

△ *White colour is a recessive gene. Two white 'copies' of the gene are needed to produce a white flower.*

Some inherited characteristics in humans are passed on, like the purple flower-colour gene in pea plants, because of a single, dominant gene.

But what exactly is a gene? Despite all the physical evidence that some kind of hereditary material is passed from generation to generation of people and of plants, no one could answer that question for nearly 100 years after Mendel carried out his experiments.

◁ *Can you roll your tongue like a girl in this picture? If you can, you carry the dominant gene for controlling the tongue muscle, and at least one of your parents will be able to roll their tongue as well.*

THE CODE OF LIFE

IN THE 1940s and 1950s, a series of discoveries established the nature and form of the hereditary material that is passed down from parent to child. It is the chemical DNA, short for deoxyribonucleic acid. It carries in a coded form the information that determines most of our physical development and all our inherited characteristics.

▷ *With a few exceptions (like red blood cells) every cell in our bodies contains a complete set of coded instructions. This means, for instance, that skin cells contain the code for digestive enzymes, but in the skin they are not active, so the enzymes are not manufactured. Only those instructions needed for a cell's specialised function are switched on.*

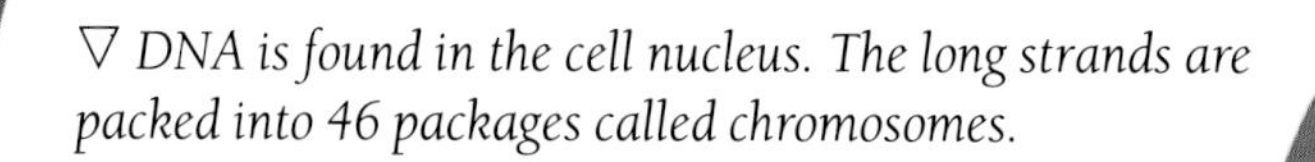

▽ *DNA is found in the cell nucleus. The long strands are packed into 46 packages called chromosomes.*

△ *If all the DNA in one cell were unravelled it would stretch for 2 m (6 ft 6 in). With about 3 billion billion cells in our bodies, each of us carries some 6 billion billion metres of DNA – enough to stretch to the Moon and back 8,000 times.*

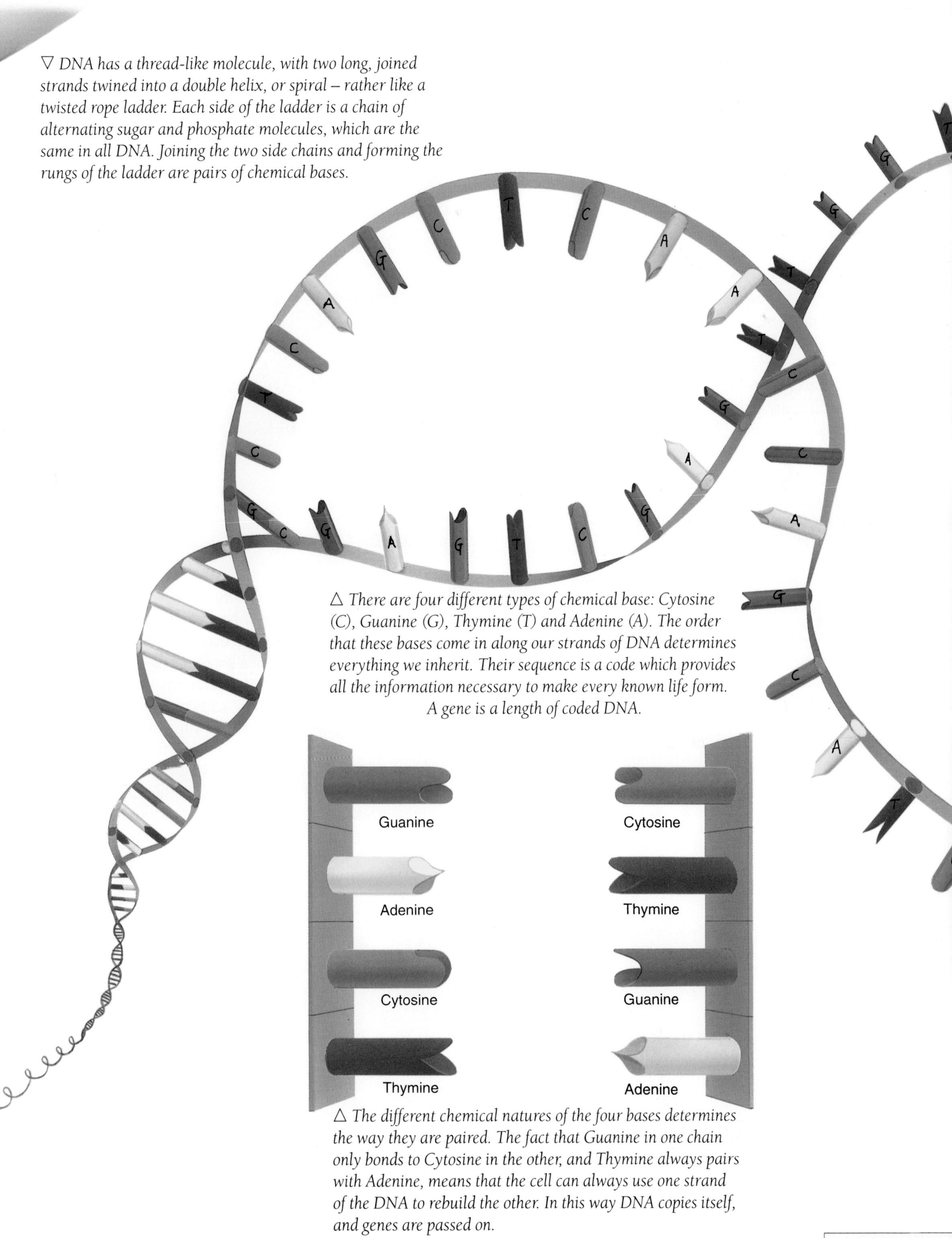

▽ DNA has a thread-like molecule, with two long, joined strands twined into a double helix, or spiral – rather like a twisted rope ladder. Each side of the ladder is a chain of alternating sugar and phosphate molecules, which are the same in all DNA. Joining the two side chains and forming the rungs of the ladder are pairs of chemical bases.

△ There are four different types of chemical base: Cytosine (C), Guanine (G), Thymine (T) and Adenine (A). The order that these bases come in along our strands of DNA determines everything we inherit. Their sequence is a code which provides all the information necessary to make every known life form. A gene is a length of coded DNA.

△ The different chemical natures of the four bases determines the way they are paired. The fact that Guanine in one chain only bonds to Cytosine in the other, and Thymine always pairs with Adenine, means that the cell can always use one strand of the DNA to rebuild the other. In this way DNA copies itself, and genes are passed on.

GENES AND PROTEINS

OUR GENES determine who and what we are through the manufacture of proteins. Each gene is a length of DNA that tells our cells how to make one – and only one – protein. Usually it takes many genes, and the interplay of many proteins, to define any one characteristic such as a smile or the shape of a nose. Geneticists estimate that it takes between 50,000 and 100,000 genes to make a whole human being.

Proteins are essential ingredients of life. They are our bodies' basic building blocks – nails, hair, muscles and skin are all made of proteins. The chemical reactions which produce energy, heat and movement, and which give us life, are governed by proteins called enzymes and hormones. Proteins called antibodies are made by our immune systems to protect us from disease.

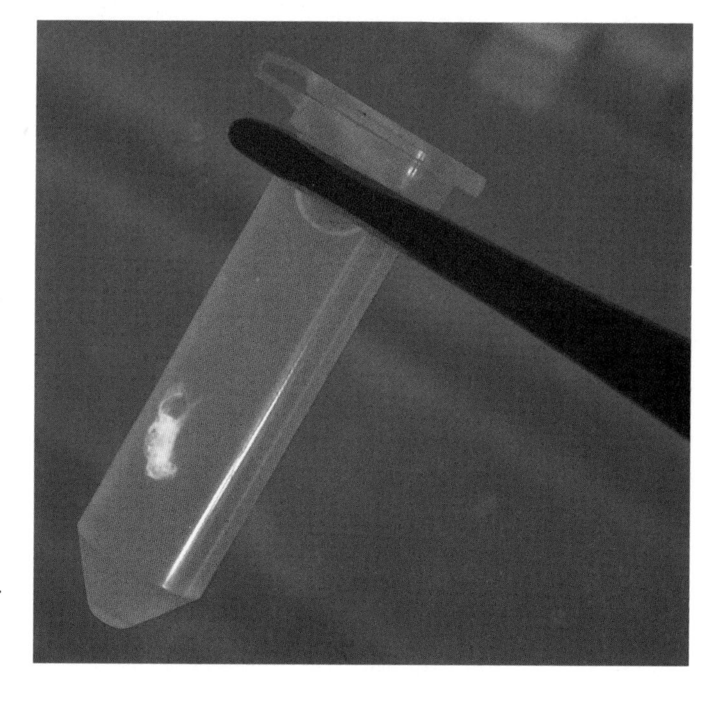

▷ *The tangled wisp at the bottom of this tube is human DNA. It carries all the instructions required to make a particular human being.*

Protein manufacture begins with the DNA in the cell nucleus. Each group of three bases along the DNA chain, known as a triplet, instructs the cell to make one particular building block of a protein called an amino acid. The genetic code runs in sequence along the DNA molecule. The series of amino acids to which it corresponds join up into a long chain, which twists and coils to form a protein. A gene is the string of bases that codes for the amino acids necessary for one protein. Every gene is several thousand bases long. Each protein contains a few hundred amino acid building blocks. But although there are many thousands of different proteins, there are only 20 different types of amino acid from which they are all constructed.

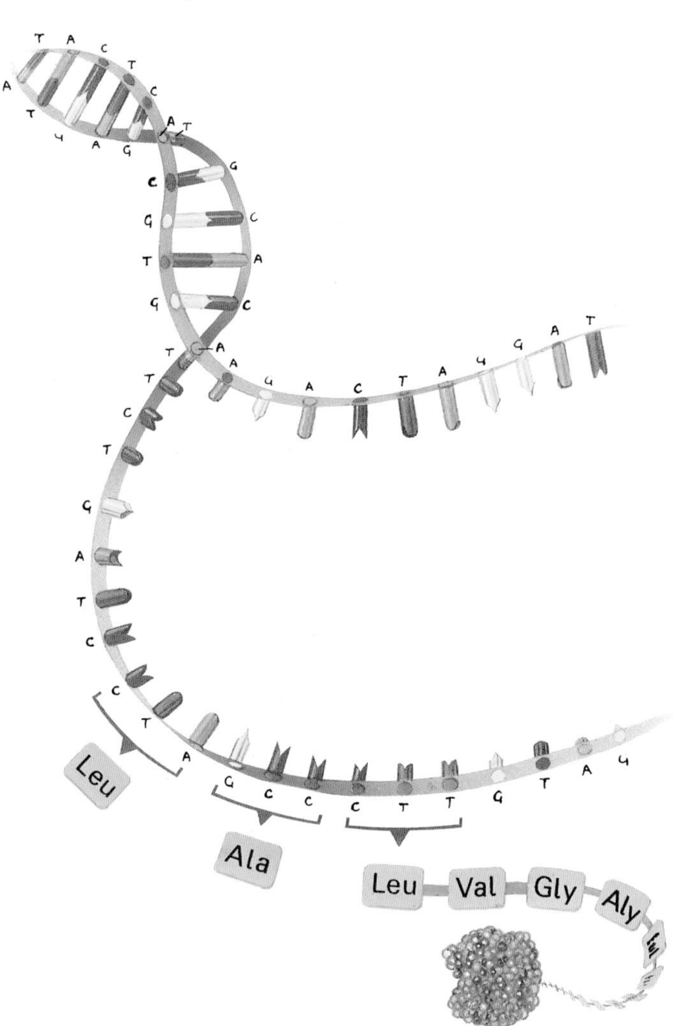

◁ *Using intricate biological machinery the cell translates DNA triplets into amino acids and joins them together to make a protein, in this case trypsin.*

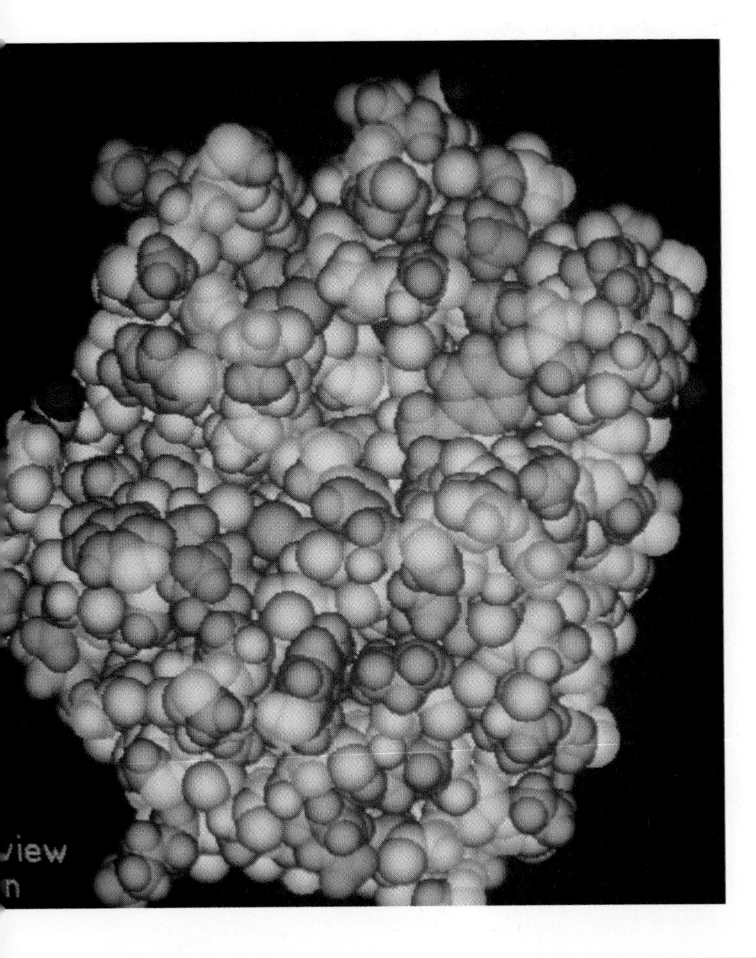

Since there are four bases, there are $4 \times 4 \times 4 = 64$ possible ways of making up a triplet, more than enough to act as a code for the 20 amino acids. Indeed, most amino acids are coded for by more than one triplet. For example, the sequence of the three bases Guanine, Cytosine, Cytosine (GCC), codes for the production of Alanine, and so does the triplet GCT.

Genes which code for proteins are known as structural genes. There are stretches of DNA that act as switches that turn the structural genes on and off. In human cells only those genes needed for that cell's specialised function are switched on. The code for methionine (ATG) acts as a signal for the cell to start protein production. The three triplets TAA, TAG and TGA are stop signals.

Genes themselves make up only about 3 per cent of our DNA. The remaining 97 per cent apparently does nothing and is known as 'junk' DNA. But it might not be 'junk' at all. As creatures evolve, 'junk' DNA could act as source material for new genes. When a mutation changes the order of bases in a gene, a new gene which codes for a different protein is created but the original gene is lost. Mutations in this apparently spare DNA would form new genes without losing anything. The more complicated an organism is, the more genes it has – and they have to come from somewhere.

△ *This is a computer graphic representation of a molecule of the protein trypsin. Trypsin is an enzyme which help us to digest the proteins in our food.*

These are the commonly used abbreviations for the 20 amino acids coded for by DNA.

Ala	Alanine	Leu	Leucine
Arg	Arginine	Lys	Lysine
Asn	Asparagine	Met	Methionine
Asp	Aspartic acid	Phe	Phenylalanine
Cys	Cysteine	Pro	Proline
Gln	Glutamine	Ser	Serine
Glu	Glutamic acid	Thr	Threonine
Gly	Glycine	Trp	Tryptophan
His	Histidine	Tyr	Tyrosine
Ile	Isoleucine	Val	Valine

1st position ↓	2nd position				3rd position ↓
	T	C	A	G	
Thymine T	Phe	Ser	Tyr	Cys	T
	Phe	Ser	Tyr	Cys	C
	Leu	Ser	STOP	STOP	A
	Leu	Ser	STOP	Trp	G
Cytosine C	Leu	Pro	His	Arg	T
	Leu	Pro	His	Arg	C
	Leu	Pro	Gln	Arg	A
	Leu	Pro	Gln	Arg	G
Adenine A	Ile	Thr	Asn	Ser	T
	Ile	Thr	Asn	Ser	C
	Ile	Thr	Lys	Arg	A
	Met/START	Thr	Lys	Arg	G
Guanine G	Val	Ala	Asp	Gly	T
	Val	Ala	Asp	Gly	C
	Val	Ala	Glu	Gly	A
	Val	Ala	Glu	Gly	G

▷ *The genetic code. Following the blue stripes from G on the left, C at the top and C on the right to where they cross at Alanine shows how the code table works.*

GOOD THINGS IN SMALL PACKAGES

THE DNA in our cells is tightly coiled and packaged into structures called chromosomes. Every time a cell divides, its chromosomes are duplicated, so that each new cell carries the same genetic information. Chromosomes are most easily seen just before cell division, when they are short, thick and X-shaped.

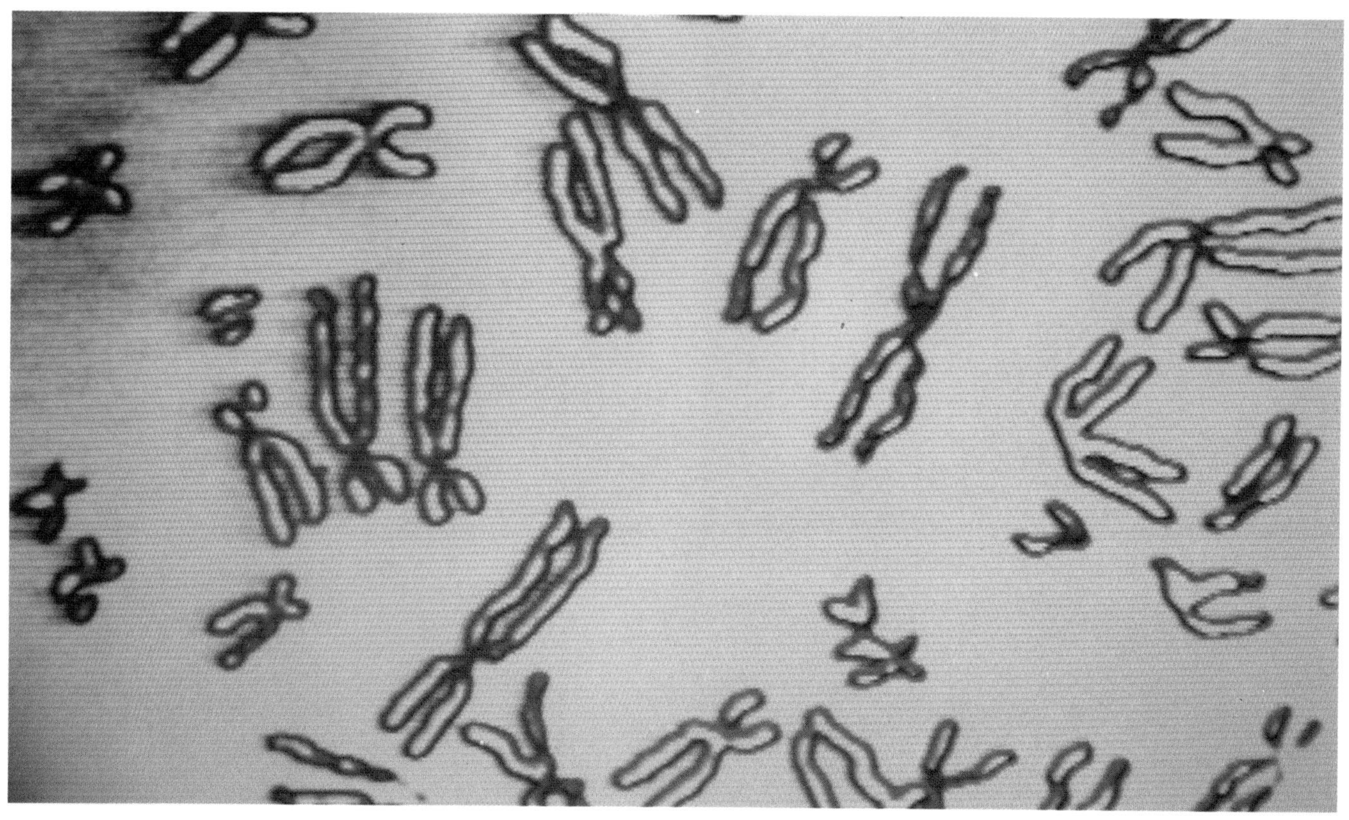

△ *This micrograph shows X-shaped double copies of single human chromosomes.*

It is possible to tell the difference between chromosomes by measuring their length. They also exhibit unique patterns of dark and light bands when stained with certain dyes. The complete set of chromosomes specifies all our inherited characteristics.

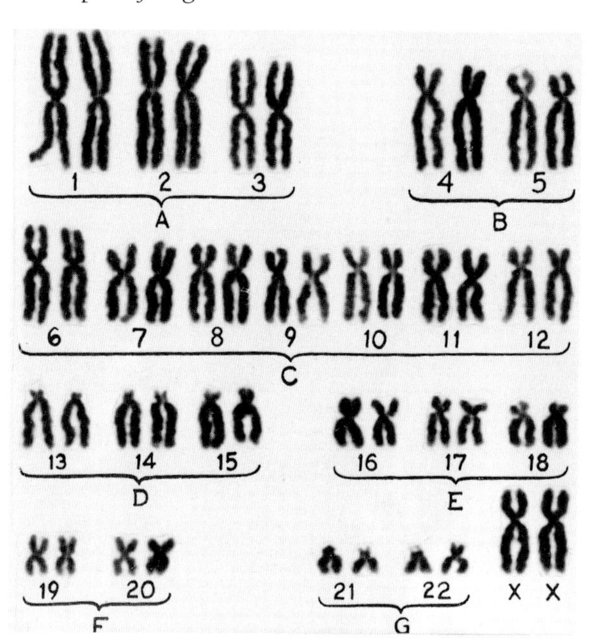

▷ *This is a full set of normal female chromosomes. The 23rd pair marked 'X' shows this is a woman (see page 14).*

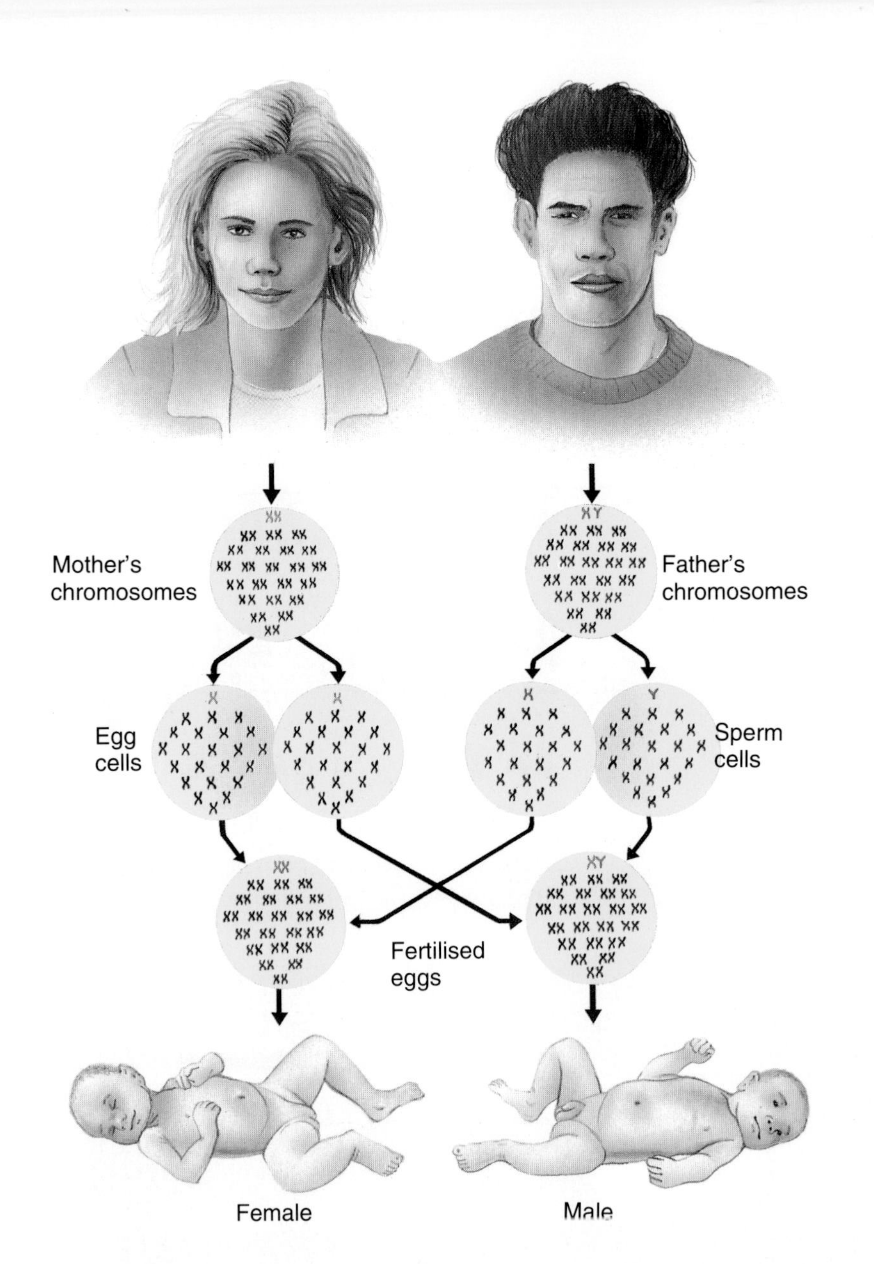

All our cells hold 23 pairs of chromosomes. One of each pair was inherited from our mother, and one from our father. Eggs and sperm carry only half this amount, 23 single chromosomes. During the formation of egg and sperm, genes are rearranged within chromosome pairs so each egg and sperm ends up with 23 chromosomes that hold new combinations of genes.

When egg and sperm combine, the resulting embryo has 23 chromosome pairs containing a unique combination of the parents' genes. This shuffling of genetic information every time an egg or a sperm is made, and every time they join together, means that any couple can bear an almost infinite number of different children.

◁ *Body cells contain 23 pairs of chromosomes. Eggs and sperm have only 23 single chromosomes. When egg and sperm combine, a cell with a full set of 23 pairs of chromosomes is created. This fertilized egg can develop into an individual with a new combination of chromosomes and a unique set of inherited characteristics.*

Two Californian geneticists, working with the repeating and interweaving patterns of the genetic code, became convinced that they were constructed along the same principles that govern musical composition. So they allocated notes to each of the four bases: Cytosine = do, Adenosine = re and me, Guanine = fa and so, Thymine = la and ti. The geneticists then transcribed some genes.

Below is the musical transformation of part of a gene from a black, inbred strain of mice. The geneticist cum composer likens it to a Chopin Nocturne. He says the more ancient a creature's origins, the more complicated its DNA, and the richer the music it makes; relatively recent DNA leads to music that is much less precisely structured, and more like Debussy or Ravel.

This work may sound strange and not very useful, but the researchers say that hearing genes as music helps them to remember the base sequences and to spot a familiar sequence if it crops up again in a different gene.

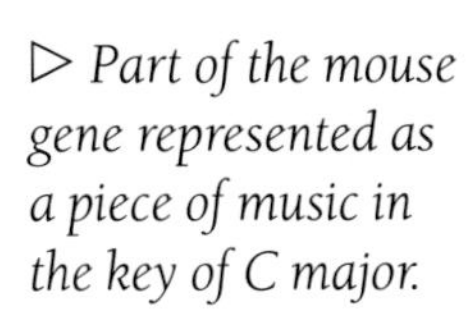

▷ *Part of the mouse gene represented as a piece of music in the key of C major.*

THE BASIS FOR SEX

WHAT SEX you are is determined by one particular pair of chromosomes, known as the X and Y chromosomes.

This set of chromosomes looks, at first sight, just like the set on p.12, but there is one crucial difference. The earlier set belongs to a woman and contains two X chromosomes. This set belongs to a man, and has only one X, and one much smaller chromosome, known as the Y chromosome. The X and Y chromosomes are known as the sex chromosomes.

While many different genes are probably required to determine our sex, embryos develop as females unless the genes on the Y chromosome are switched on. Geneticists have long been searching for the single gene on the Y chromosome that acts as the switch for male development.

△ *This is a full set of 23 pairs of human male chromosomes. The two X chromosomes from a female are replaced with one X and one Y.*

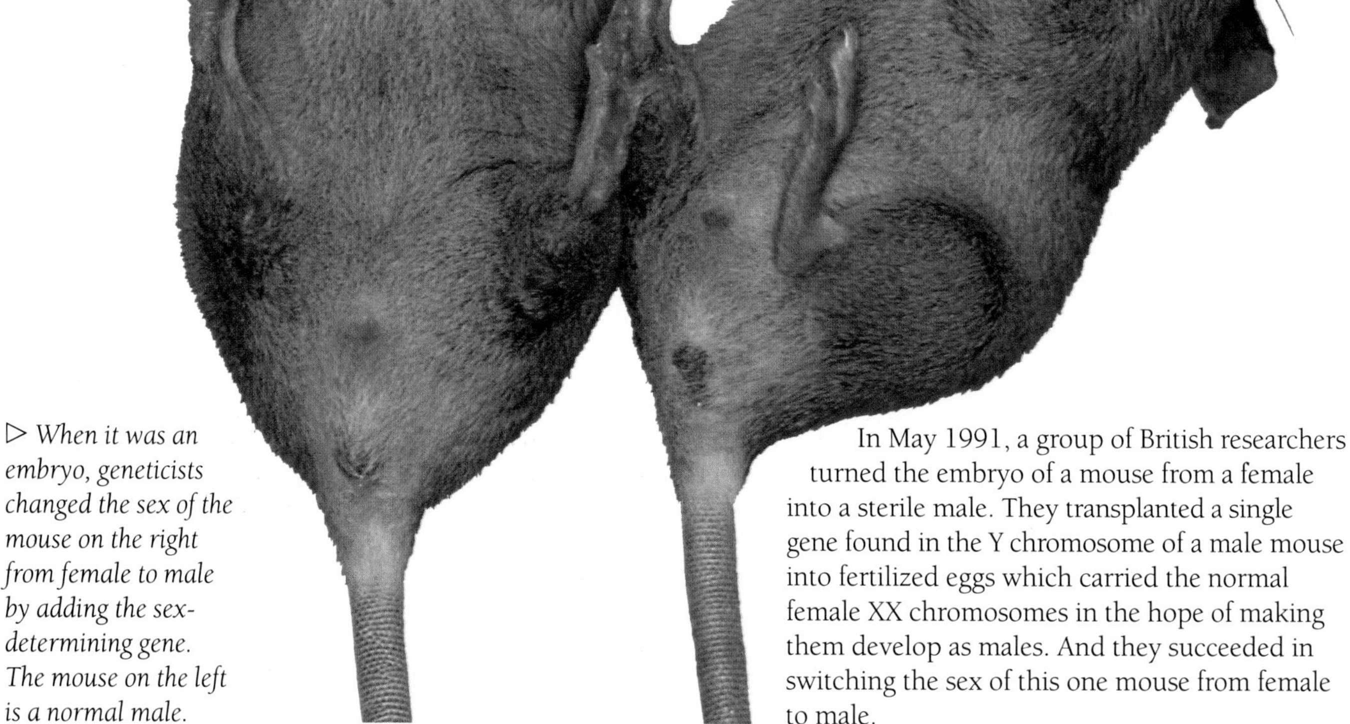

▷ *When it was an embryo, geneticists changed the sex of the mouse on the right from female to male by adding the sex-determining gene. The mouse on the left is a normal male.*

In May 1991, a group of British researchers turned the embryo of a mouse from a female into a sterile male. They transplanted a single gene found in the Y chromosome of a male mouse into fertilized eggs which carried the normal female XX chromosomes in the hope of making them develop as males. And they succeeded in switching the sex of this one mouse from female to male.

Since the early 1960s, every female athlete has had to undergo a sex test before she can enter important international competitions, such as the Olympics, as a woman. In 1968, genetic tests were introduced, based on the idea that only two X chromosomes can make someone a woman. The trouble is that this is not always true.

Most of the people disqualified by these tests have been women with unusual chromosomal conditions. For instance, some people are born with a normal male chromosomal make-up, but either their Y chromosome genes are not switched on and they do not produce male hormones, or their cells cannot respond to the hormones they do produce. So no male hormonal messages get through, and these individuals develop physically as normal women.

In 1992, the International Amateur Athletics Federation abandoned genetic testing. They decided that a medical examination which includes a simple inspection of the sex organs should confirm both the good health of the athlete, and their sex.

The male Y chromosome carries only genes that relate to sex, whereas the X chromosome carries many that have nothing to do with sex determination. Having only one copy of the X chromosome leads men to suffer from a group of inherited diseases that, on the whole, do not affect women.

△ *The women's 4 × 100 metre relay in the Seoul Olympics, 1988. At that time, women athletes underwent genetic sex tests.*

When abnormal genes on the X chromosome are dominant, they show their effects in both sexes, but recessive abnormalities are different. With two X chromosomes, women with a recessive abnormality on one usually have a normal gene on the other and this keeps things working well. In men, however, where there is no second X chromosome, any abnormality on the single X chromosome can lead to a problem.

Best known of the recessive, X-linked disorders are colour blindness and haemophilia. Both conditions are carried by normal females but they only appear in those sons who inherit the abnormal genes. It is only possible for a woman to be colour blind if she inherits two copies of the deformed gene, one from an affected father and one from a carrier mother – a very rare combination.

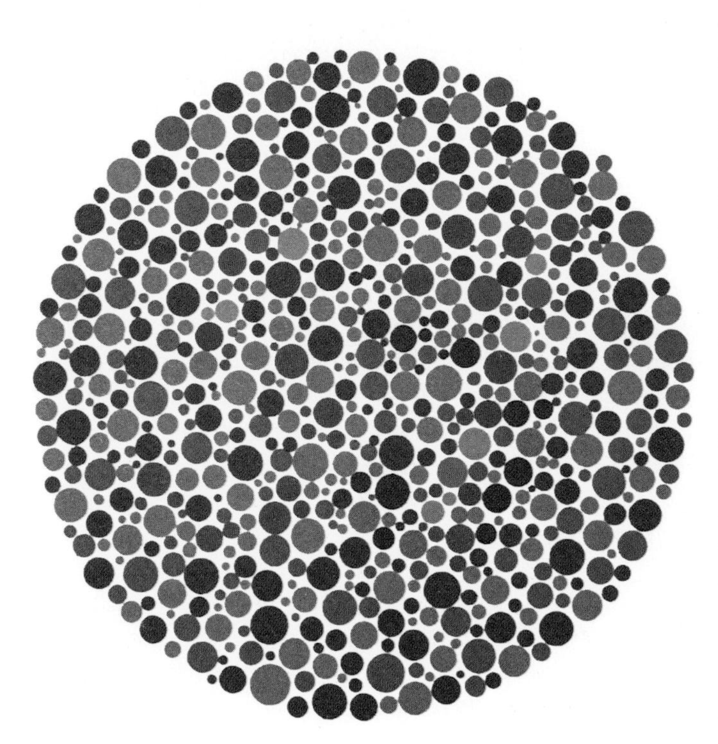

◁ *You should be able to read the number 57 in this pattern of dots. If you are red-green colour blind you will see the number 35.*

THE NEW ENGINEERING

THE DNA that genes are made from is the same in every living thing, whether hyacinth, hippopotamus or human. In the mid-1970s, a group of bacterial enzymes was discovered that can be used to snip pieces of DNA out of one set of genes and splice them into another. This made possible the reliable transfer of genes between organisms. Now genetic engineers routinely move genes around, even from animals to plants, creating cells with properties that have never existed before.

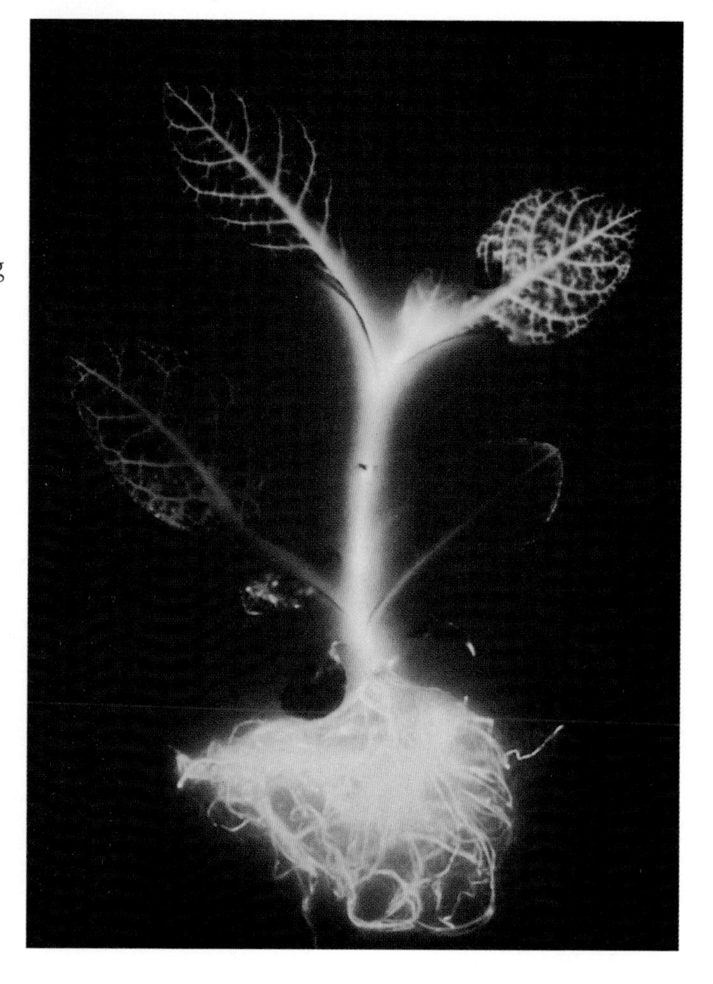

▷ *This tobacco plant has had the gene for luciferase, the chemical that makes fireflies glow in the dark, engineered into its genetic material.*

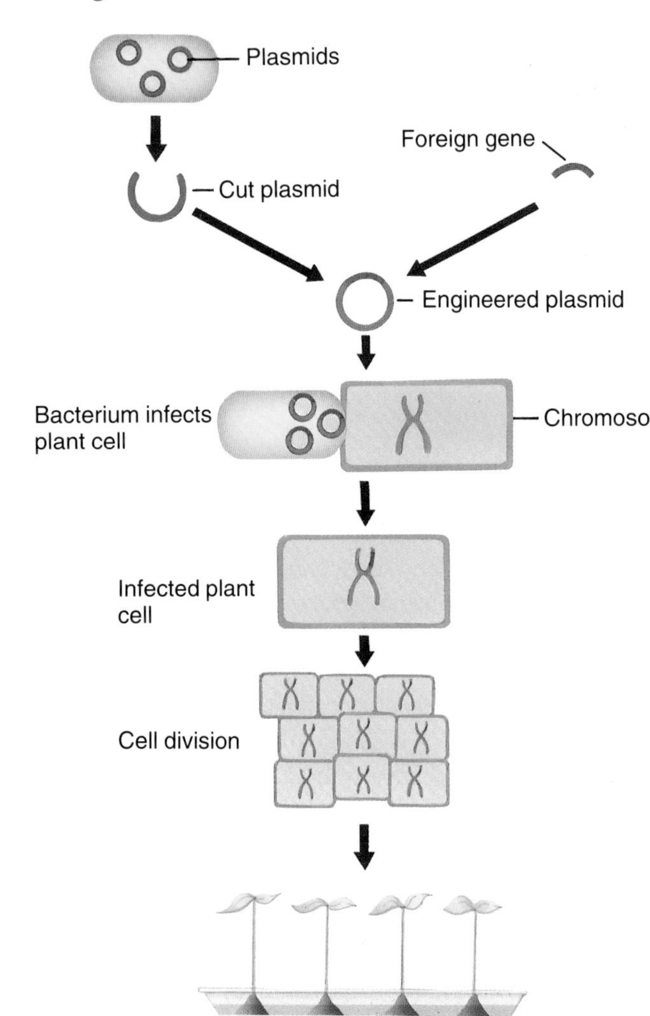

To engineer the genes of a plant, bacteria that infect it naturally in the wild are used. Inside bacteria there are small loops of DNA called plasmids. Through intricate genetic surgery, using bacterial enzymes as scissors, foreign genes can be put into the plasmids. These genes come from plants or animals, and code for the characteristics required. The modified bacteria are then used to infect cells in plant material.

When bacteria infect a plant, they naturally insert their DNA plasmids into the plant's own genes. Because the changed bacteria cannot distinguish the foreign DNA from their own, the foreign genes are inserted into the plant's genetic material along with the rest of the plasmid. Plants that are then grown from the infected leaf produce the foreign protein and have the characteristics for which it is responsible.

◁ *Insect or herbicide resistance can be introduced into crops that need it, by transplanting the genes that code for protective proteins from naturally resistant plants or even bacteria.*

One of the most promising characteristics gene transfer has to offer is resistance to pests and disease.

The cotton plant on the left has been modified to produce a chemical that kills caterpillars, including weevil larvae. The foreign gene comes from a bacterium that produces the insecticide naturally, but it has been redesigned by scientists in the laboratory to make it even more effective against bol weevils. This work could halve the use of insecticides on cotton.

▷ *Both these plants were exposed to bol weevils. The plant which had been genetically engineered to produce an insecticide grew healthy, fluffy cotton bols.*

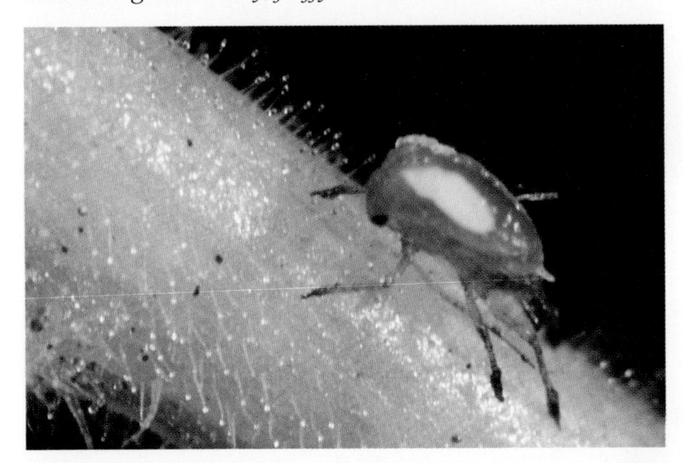

△ *An intrepid beetle takes on the stalk of a sticky potato. This could well be its last meal.*

Sticky hairs keep some potatoes pest free in a very different way. Geneticists in Peru spent ten years combining genes from a wild, hairy, non-edible strain of potato and a tasty domestic variety. Small insects stick to the hairy leaves and stalks. Larger pests, like the Colorado beetle, eat the hairy vegetation and gum up their insides with the sticky goo that tips each hair.

Genetic engineers are making many crops resistant to viral diseases and tolerant of common herbicides. Resistance to drought and salt, though, depends on large groups of genes, so introducing it to crops like rice and wheat is more complex. Work like this, and other research aimed at adding protein to staple crops like corn and cassava, might one day help the world to feed its growing population.

There are plans afoot to make tomatoes frost resistant by introducing into their DNA a gene from the Arctic flounder. However the first genetically engineered food to go on sale to the public, is a tomato which contains no additional genes at all – one of the genes has been switched off.

Tomatoes ripen and rot faster than they can be brought to market. So they are usually picked before they are ripe, and transported while still green and firm. Then they are sprayed with a chemical that makes them turn red, and sold. Since flavour comes as tomatoes ripen on the vine, this means that most of the tomatoes we buy are firm but tasteless. By inactivating the gene that controls the rotting process, agricultural genetic engineers have created a tomato that will ripen slowly on the vine and develop its full flavour, but will nevertheless stay firm enough to stand the rigours of transportation.

▷ *More than 40 different crops and vegetables are having their natural qualities altered in research laboratories around the world.*

BUGS AND DRUGS

MANY IMPORTANT drugs are proteins but they are difficult and very expensive to make, a fact which sometimes limits their use. Now, genetic engineering is making hundreds of pharmaceuticals available in quantities and at prices not possible before.

Since the beginning of the 1980s, bacteria with human genes engineered into their plasmids have been put to work on an industrial scale. Given the right nutrients, warmth and moisture, they reproduce as normal, doubling their numbers every twenty minutes, and produce the human protein as they go. In 1982, insulin became the first common drug to be made by genetically engineered bacteria. Before that, it had always been extracted from the pancreases of pigs and cows.

Many proteins are too large and complicated for bacteria to make. So the gene 'pharmers' are turning to big animals which can be made to produce pharmaceuticals in their blood and milk.

△ *Technicians must wear clean room clothing so they do not contaminate the fermentation vats and the millions of bacteria they hold. All the bacteria produce the same protein, which is separated and purified before use.*

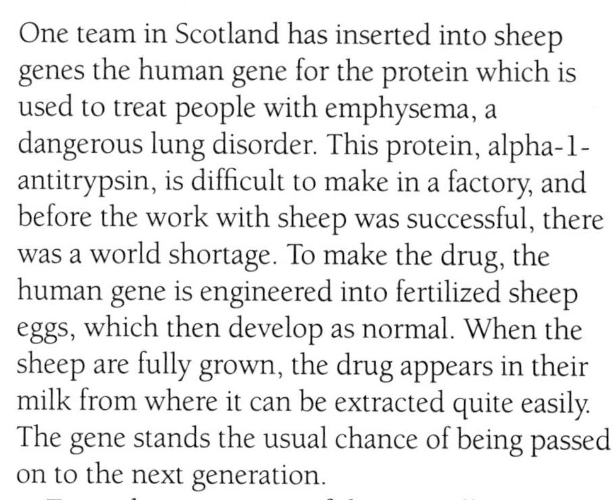

One team in Scotland has inserted into sheep genes the human gene for the protein which is used to treat people with emphysema, a dangerous lung disorder. This protein, alpha-1-antitrypsin, is difficult to make in a factory, and before the work with sheep was successful, there was a world shortage. To make the drug, the human gene is engineered into fertilized sheep eggs, which then develop as normal. When the sheep are fully grown, the drug appears in their milk from where it can be extracted quite easily. The gene stands the usual chance of being passed on to the next generation.

Tracy, the most successful genetically engineered sheep so far, produces 70 grammes of alpha-1-antitrypsin in every litre of milk. Made conventionally, the drug costs $100 per gramme, so Tracy is a very valuable sheep. A flock of just 1,000 sheep like her would currently match the entire world production of the protein. Eventually, whole herds of Tracys or even cows will be put to work as commercial bioreactors.

◁ *Tracy and two of her daughters. The lamb on the right inherited the human gene from Tracy and will reproduce the valuable drug.*

Plants are even cheaper to breed than bacteria or animals. If drug manufacturers can get even tiny amounts of a drug into a plant, then they will be able to mass produce it in large fields of the crop. Exploiting this idea still depends on developing ways of separating the drug from the rest of the crop, but there are already teams trying to make monoclonal antibodies and insulin in the leaves of crops like tobacco and alfalfa.

Drugs are not the only surprise waiting in the fields. The plastic that this bottle is made from comes from rather unusual bacteria. These bacteria use plastic as an energy store, in much the same way as a hibernating animal lays down fat. Now the same biodegradable plastic can be made in the leaves of a common weed.

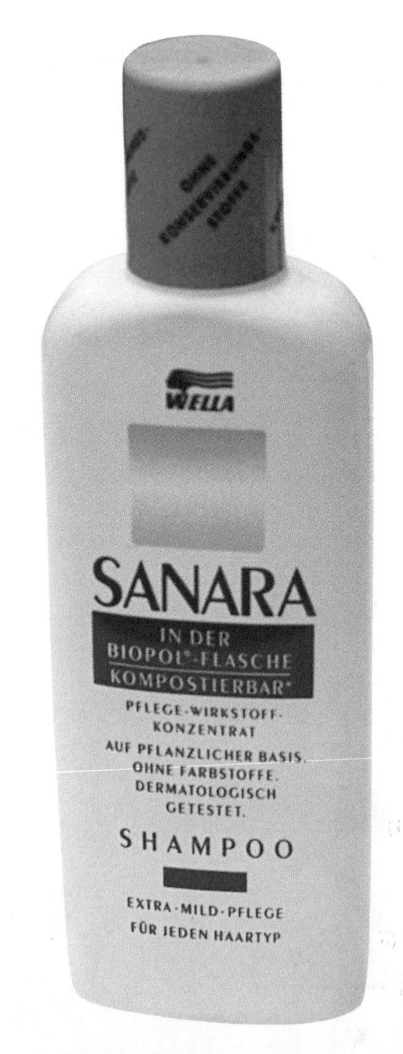

▷ *This German shampoo bottle is made from plastic produced by bacteria.*

Thale cress has had the genes that code for the biodegradable plastic transferred to it from the bacteria, and as the plant grows the plastic appears as microscopic beads in its leaves. But thale cress needs the carbon the plastic is made from to stay alive. Diverting its carbon stores into plastic production makes the plant unhealthy. The researchers expect to have more success by transferring the gene into potatoes or sugar beet. These both produce vast amounts of carbon-rich starch and sugar and could produce a practical plastic crop.

△ *Research carried out on thale cress could herald a new era of plastics produced by plants.*

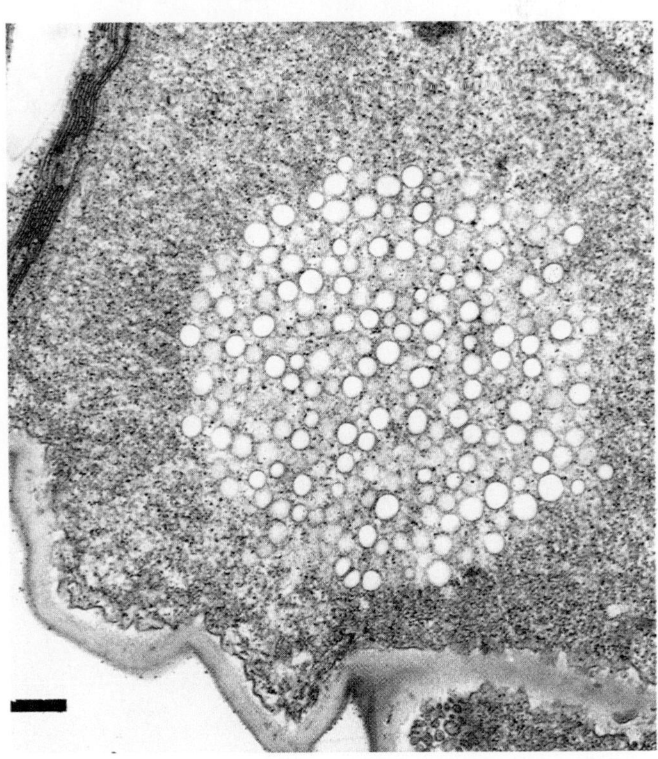

△ *Circular plastic granules collect in the cells of genetically engineered thale cress plants.*

PROTESTS AND PATENTS

THE GENETIC engineering of plants and animals offers the world astounding possibilities. However, while, in general, geneticists are trying to make improvements in food and medical care, many people question the motives of some of those involved, and indeed question the need for any genetic interference.

One concern is that not enough is yet understood about how living things interact for us safely to release genetically altered creatures and plants into the environment. We already know that plants exchange genes quite naturally. What would happen if a gene engineered into a crop, perhaps to give it resistance against pests, crossed into a weed?

Biotechnology companies claim that genetically engineered crops will reduce our dependency on chemical pesticides and fertilizers. This may be true, but a huge number of plant varieties are also being modified so that they can survive when fields are sprayed with a particular herbicide. The idea is then to sell them as a package, seeds and herbicide together. While the herbicide concerned may be better for the environment than those that are currently in use, this practice will tie farmers to one brand and the main beneficiary might be the controlling company.

If resistant varieties of crops become popular, ironically our vulnerability to pests against which resistance has not been engineered will increase. In 1970, in the USA, a virulent new strain of fungus killed half the maize crops between Miami and Texas, causing $1 billion worth of damage. This happened because farmers were growing one basic variety of maize and it could happen again if farmers all began planting identical engineered crops.

△ *Many people feel uneasy about the way genetic engineering is being ever more widely used to alter living things.*

▷ *Genetic engineering of crops could reduce the need to spray against weeds and pests.*

The ability to manipulate the genes of animals is creating greater ethical dilemmas. Even among those who agree with genetic manipulation in principle there are disputes, not least over the question of whether anyone should be able to 'own' life.

In 1988, the US patent office granted the world's first patent on a mammal. It was given to Harvard University, for a mouse tailor-made for cancer research. The 'OncoMouse', which is now in the process of being patented in Europe, was genetically engineered so that it and its offspring are more likely than normal mice to develop tumours. But the furious debate that followed the granting of this patent has stalled all other, similar applications in Europe and the USA.

▽ *Mouse number 4,736,866 was the first mammal in the world to be patented.*

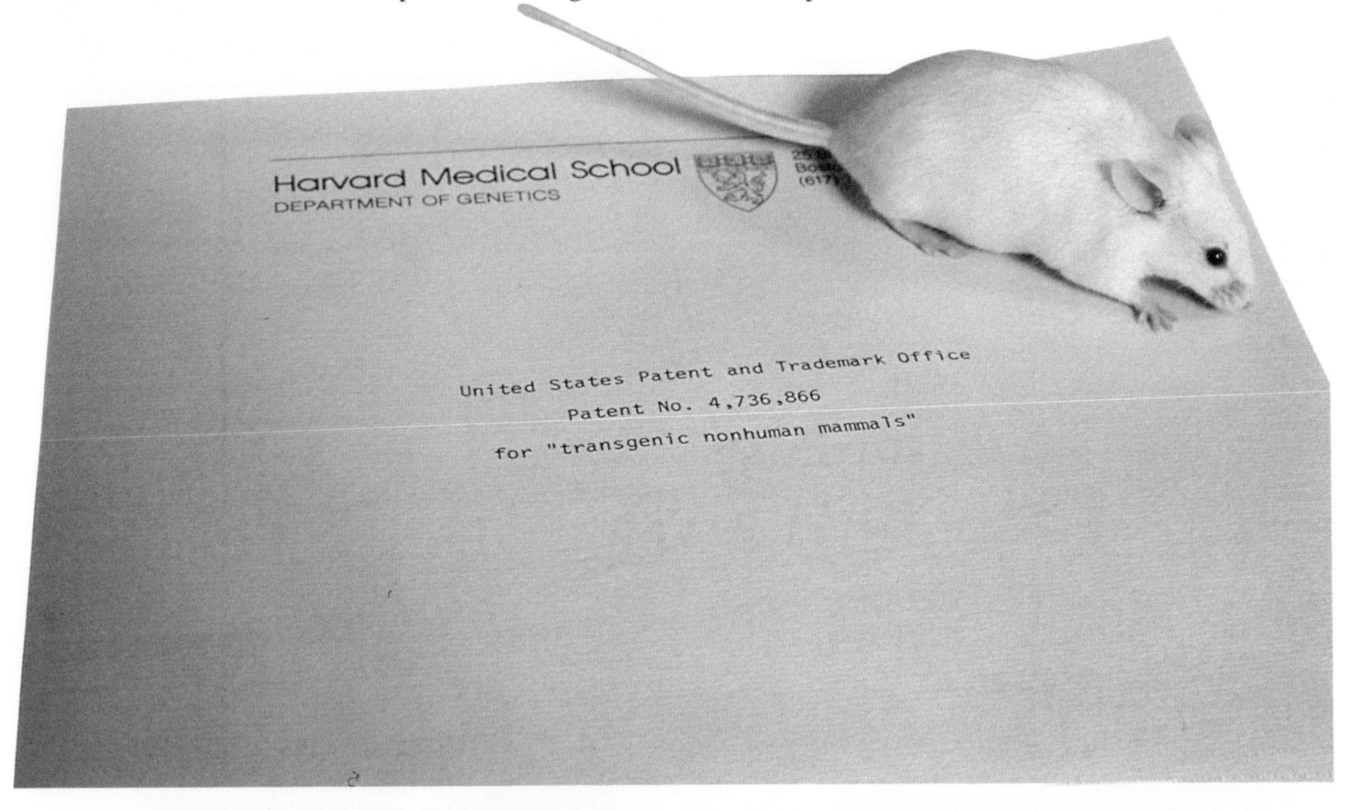

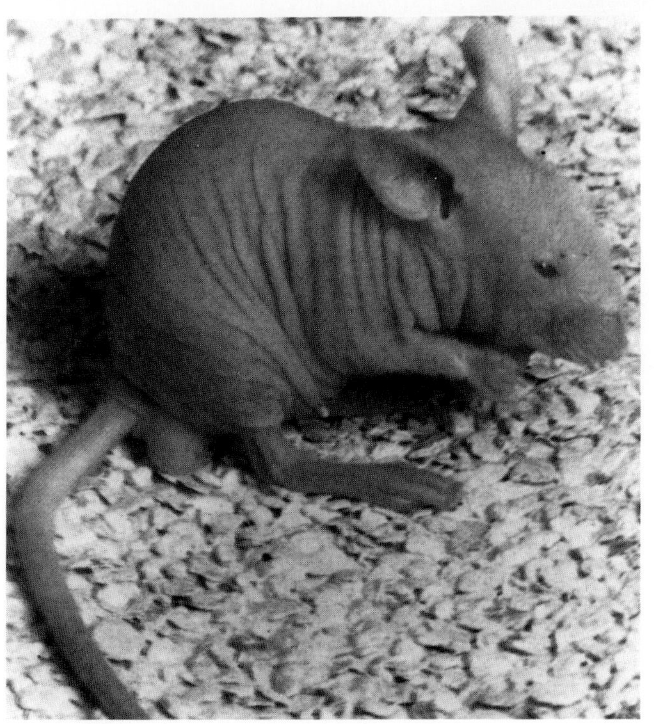

An American pharmaceuticals company had its application for a patent on a hairless mouse turned down in Europe and in the USA on the grounds that it involved suffering for the animal that far outweighed the benefit to people.

And it is not just whole organisms that are being claimed. As the study of human genes progresses, some geneticists are trying to patent the base sequences they discover, even though in many cases they have no idea what they do. This approach to genetic research could destroy international collaboration and hinder product development. A partial solution to the problem would be for the patent applications to be altered so that they cover the specific uses to which genes and gene fragments would be put, rather than the very structure of the DNA. But this will not satisfy those who dispute that any one person or organization should own the rights to something so fundamental to us all.

◁ *A patent was refused on this strain of hairless mouse, used to test hair restorers.*

SILENCE OF THE GRAVE?

GENES ARE joining fossils as a major tool in the unravelling of evolution. In the same way as fossil hunters piece together the skeletons of extinct species, genetic "archaeologists" are trying to assemble the genetic material of ancient creatures and people. DNA has been extracted from samples of blood, organic tissues and even dry bones that are many thousands of years old. The genetic patterns that are revealed are beginning to shed new light on the nature and movements of people and animals in history.

The oldest human DNA so far recovered came from bodies buried in Florida between 7,000 and 8,000 years ago. Archaeologists excavated 177 skeletons and 91 well-preserved brains from the site. DNA sequences from genes that regulate the immune system have been recognized in tissue from the brains. They may eventually reveal the diseases from which these early Americans suffered and to which they were resistant. The genetics of the population seem to have changed little over the 1,000 years the burial grounds were used, which implies that the community was very inbred.

Investigations of ancient DNA rely on minute bits of genetic material. At first, many scientists did not believe DNA could survive for even hundreds of years and worried that supposedly old DNA had actually come from modern contamination, for example from the researcher's skin. The dispute was laid to rest when a 400-year-old pig bone, salvaged from the wreck of Henry VIII's battleship, the *Mary Rose*, was studied, and the DNA sequences extracted were found to be definitely those of a pig.

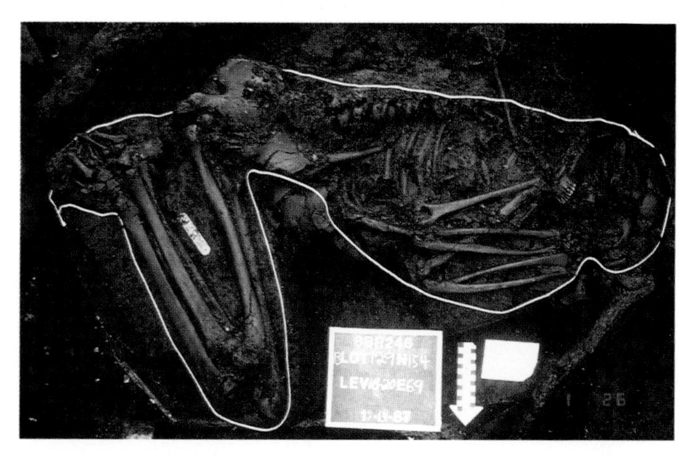

△ *A native American burial site in Florida yielded the oldest known human DNA.*

A weevil holds the world record for the oldest genetic material yet known to have survived. It was caught some 130 million years ago in a drop of sticky sap from a prehistoric tree. When sap fossilizes, it turns into amber which preserves organic tissue very well, and the insect has yielded short, recognizable stretches of DNA. The researchers hope to use DNA from the weevil and other amberized insects to answer many questions about why some kinds of insects survived while others died out.

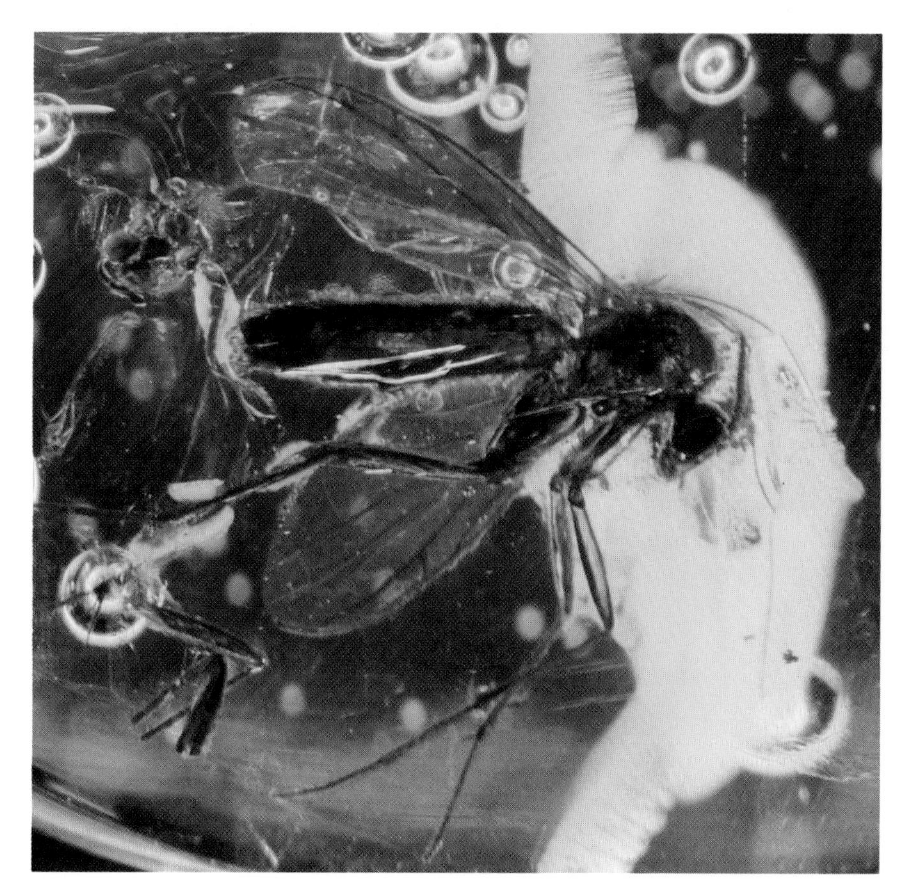

◁ *This gnat was 'sapped' 40 million years ago, and has been perfectly preserved in amber.*

Some amber specimens from 80 million years ago hold mosquitoes, with their guts still full of blood. George Poinar, a Californian geneticist, is hoping to isolate genetic material from these tiny, ancient blood samples, without knowing what kind of creatures the blood came from. Given the age of the amber samples, it is possible that he could find it is DNA from a tyrannosaurus rex or a triceratops.

George Poinar's early work inspired the novel and film *Jurassic Park*. In the story, an ambitious scientist recreates dinosaurs from DNA extracted from just such blood-sucking insects encased in amber. The book is pure fiction, and most scientists believe the recreation of dinosaurs is still completely impossible. Though the DNA from amber specimens is in better condition than fragments extracted from other fossils, it has still been terribly broken and altered by the passage of time. So the chances of piecing together all, or even most of, the genes of an ancient creature are extremely slim.

▷ *The tyrannosaurus rex from Jurassic Park is the closest we will ever come to a real dinosaur. Although dinosaur DNA may well be recovered, it is very unlikely that it could ever be used to reconstruct the animal.*

One creature that is working its way back from extinction is the quagga, a curious hybrid of horse and zebra, with stripes at the front, but not at the back. The last known quagga died in Amsterdam Zoo in 1883, but scientists in South Africa, where the creatures used to live, are trying to resurrect the animal.

In 1983, genetic material taken from the dried tissue and blood of stuffed quaggas was matched with that of the still living plains zebra. The close match meant that the quagga was not a distinct species, and that it is not truly extinct. Since then, selective breeding between modern zebras with faint stripes has produced at least one foal that looks just like a quagga should.

△ *This young plains zebra is the nearest thing to a resurrected species.*

▷ *Genetic material taken from stuffed quaggas has been shown to be very similar to the DNA of the still living plains zebra.*

THE GENETIC ADAM & EVE

DID THE MOTHER of us all live in Africa some 200,000 years ago? Many molecular biologists, tracing the evolutionary history of humans from changes found in their modern genes, believe that she did. Their controversial theory has upset a lot of traditional anthropologists whose work largely depends on fossils, and who believe that modern human beings had no single place of origin. But the uproar that the new theory has caused shows the power of using genes as tools for peering into prehistory.

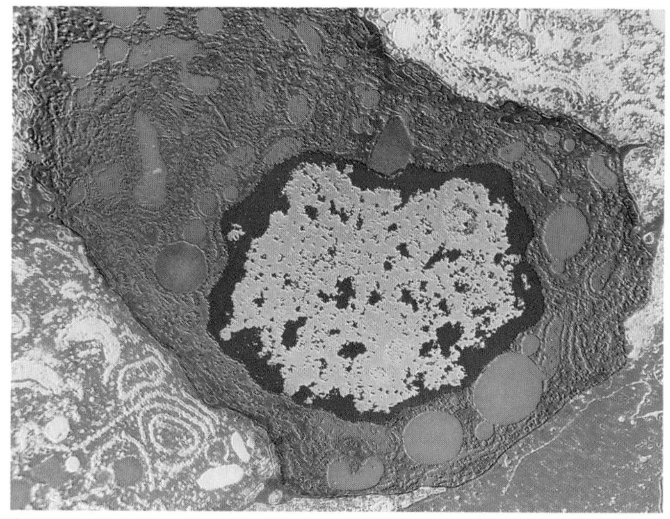

△ *In every human cell there are many mitochondria. In this electron micrograph of a human cell, they appear as elongated orange blobs.*

The cell nucleus is not the only place where DNA is found. A very small amount exists as small loops in tiny cellular structures where carbohydrates and fats are converted into energy for the rest of the cell. These little powerhouses are called mitochondria, and they are where the quest for 'Eve' began.

Mitochondria exist in the main body of the cell – one cell can hold as many as 2,000 mitochondria. In sperm, they are found in the tail, but since this snaps off during fertilization, all our mitochondria and all our mitochondrial DNA come directly from our mothers, unchanged except for chance mutations. Whereas DNA in the cell nucleus contains up to 100,000 genes, mitochondrial DNA codes for only 37, but every one is essential – a mutation in any of these genes causes severe neurological problems. But these tiny loops also contain stretches of apparently useless 'junk' DNA.

All DNA mutates from generation to generation. To track down where the human race originated, mitochondrial DNA was collected from people of African, Asian, European and Middle Eastern descent, and the pattern of mutations in certain stretches of their junk mitochondrial DNA was determined. Next a family tree was drawn up, linking the different mutation patterns. The resulting tree had two main branches, both of which led back to Africa.

However, all the variation in mutation pattern that was observed was very slight, and this led to two theories. Firstly, that modern humans are not very old – we have not collected many mutations. Secondly, all the mitochondrial DNA now spread among the world's five billion people came from one source – in other words, from one woman!

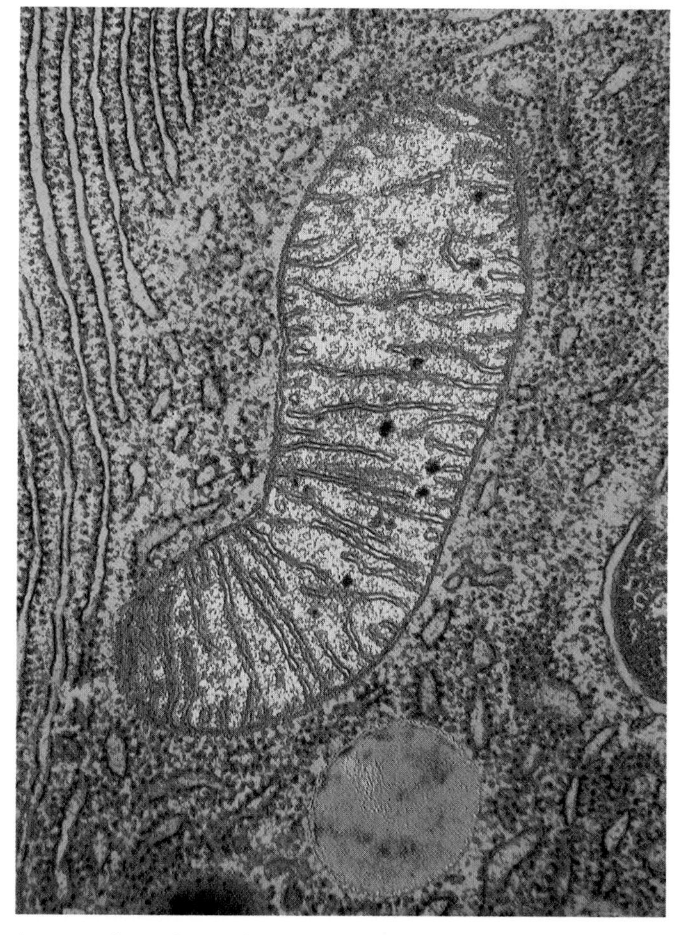

△ *Apart from the nucleus, mitochondria are the only places in a human cell where DNA is found. Their job is to combine sugars and fats with oxygen to give us energy. They are about one-thousandth of a millimetre long.*

△ *Our evolutionary history is written in our genes. It suggests the genetic Adam and Eve were African.*

How long ago did this African woman live? It has been found that mutations in 'junk' portions of mitochondrial DNA accumulate steadily, and the theory goes that the rate at which these mutations appear gives us a kind of molecular clock. The assumption that this is true, and the assumption that the clock runs at constant speed (and both points are in dispute) leads to the conclusion that one woman living in Africa somewhere between 140,000 and 280,000 years ago was the mother of us all.

To seek out an Adam for this mitochondrial Eve, researchers all around the world are turning to the Y chromosome. Since this only exists in men, it could lead back to a common 'father'. Studies of the Y chromosome are still in their infancy, but French researcher Gérard Lucotte has taken advantage of the world's first Y chromosome bank and has, he claims, identified the African tribe that our earliest common male ancestor must have come from – the Aka pygmies of west Africa.

▷ *These adult men of the Aka pygmy tribe were photographed in 1925. They are the size of an average western 11 year old.*

THE GENETIC WITNESS

DNA FINGERPRINTING is probably the greatest advance in forensic science since the advent of fingerprinting itself. It is a technique which allows forensic scientists to compare DNA from biological evidence found at the scene of a crime, with DNA from criminal suspects or perhaps from the victim of a crime. This helps the police to eliminate innocent suspects or identify the guilty party.

To make a genetic fingerprint, a sample is needed that contains some kind of biological material such as saliva, blood, semen or even hair roots. These samples contain human cells, from which the DNA for the genetic fingerprint can be extracted.

Amongst our genes there are many regions of DNA where stretches of about 20 to 40 base pairs are repeated over and over again for no apparent reason. The exact pattern of these repeats is unique to each individual. Like the rest of our genes, we inherit one version of these regions from our mother and the other from our father. Only identical twins carry the same pattern throughout their DNA. This means that these stretches of DNA can be used as genetic flags or markers to differentiate between different people. They can also be used to help geneticists to track down the genes responsible for particular diseases (see pp.32).

△ *No one else in the world shares the pattern of loops, whorls and ridges in your fingerprint.*

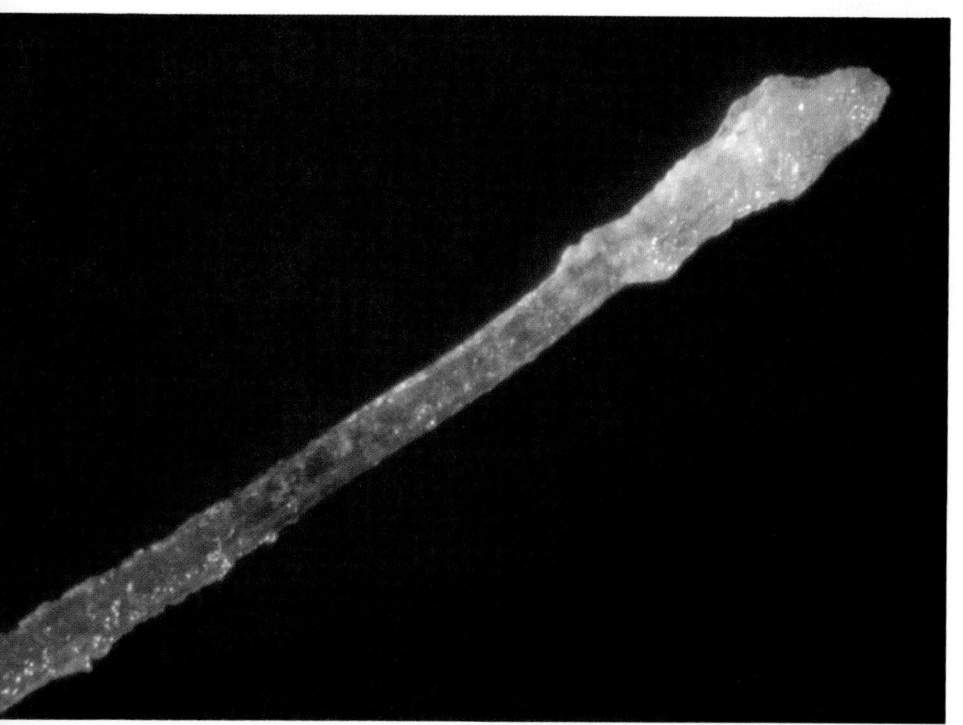

△ *A genetic fingerprint can be produced from a tiny blood sample, or from just 10 hair roots like this.*

Samples from the scene of a crime must be treated by the most sensitive method to get results from the tiniest samples of the poorest quality. Each repeating sequence, or marker, being probed shows up as a black band on a DNA strip. Since we carry two versions of each marker, every test results in two dark bands. The exact position of these bands along the strip depends on the exact number of repeats in the marker. Usually three or four different types of genetic marker are probed at the same time, which means six to eight bands per DNA strip. The particular pattern these bands make, something like a human bar code, is the DNA or genetic fingerprint of an individual.

The chance of two unrelated individuals carrying identical examples of three or four markers is perhaps as low as one in a million. The more markers used in a test, the more likely the 'fingerprint' is to identify someone.

There is, however, some dispute about just how accurate genetic fingerprinting tests are. It is argued that not enough is known about how frequently different markers appear in different populations. If some marker versions are more common, say, among Italian people than Poles, and the guilty person is Italian, then a DNA test is more likely to give a positive result with any Italian suspect than with a Polish one. Whilst these doubts are unresolved, genetic fingerprinting evidence is inadmissable in parts of the United States, but it is still accepted in most courts.

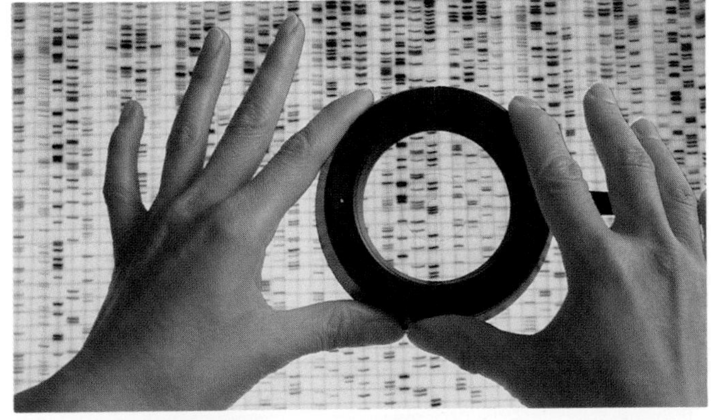

△ *These strips of dark bands are DNA fingerprints. Each band corresponds to a particular repeating sequence of DNA, or marker.*

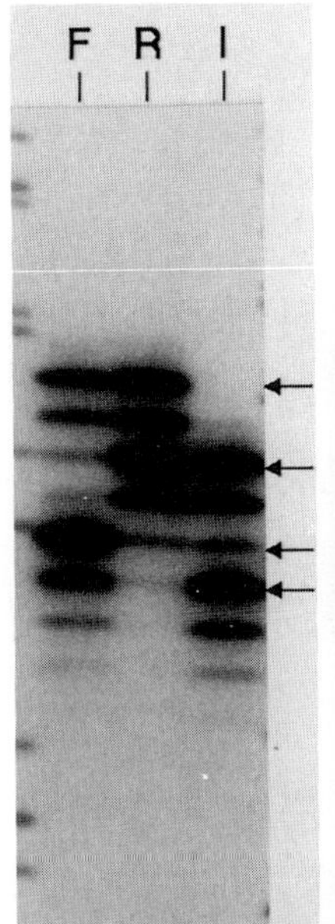

▽ *The arrows indicate marker bands in the DNA fingerprints from Mengele's femur (F) and Rolf's mother Irene (I). Rolf inherited one marker from each, suggesting the bone was Mengele's.*

Dr Josef Mengele was a doctor who devised many horrific experiments carried out in the Nazi concentration camps during the Second World War. He became known as the "Angel of Death". In 1945, he escaped to South America where in 1979 he was thought to have died of a heart attack while out swimming. However, disagreements about Mengele's known medical history and features of the skeleton found cast doubt on the identification. Comparison of DNA strips from a leg bone, a femur (F), with DNA from Mengele's wife, Irene (I) and son, Rolf (R), has resolved the dispute. By comparing the son's DNA markers with the mother's, the bands inherited from her were identified. All the remaining bands must have come from Mengele, his father.

Ten sets of bands like these were studied and in each case the paternal band in Rolf's DNA profile matched one in the bone. This added up to a less than 1 in 1,800 chance that the bone belonged to a man who was not Rolf's true father. Because of this evidence, the German authorities have closed the file on Mengele.

DNA is also used to establish paternity, both in humans and in dogs. This trainer had a problem with her Siberian huskies when the bitch Kynda unexpectedly became pregnant. In order to claim pedigree for the puppies, she had a DNA test done to establish which of her two dogs was the father.

▷ *Genetic finger-printing showed that the dog, Ali (top row, left), was the proud father of pups born to Kynda (top row, centre).*

ERRORS AND DISEASES

THERE ARE 4,000 known inherited diseases. Each one is caused by the disruption of the normal base sequence of a gene which causes the production of a wrong or faulty protein. Normally, only rare diseases are caused by a fault in a single gene, but it is believed that many common health problems such as diabetes, heart disease and even lung cancer, result from a mixture of faulty genes and the effects of the environment. Each faulty gene makes a small contribution to a predisposition in the person, but what triggers the onset of disease is an environmental factor, such as a bad diet, smoking or exposure to certain food or viruses. Someone without the same predisposition may be exposed to the same environment but never develop the condition.

◁ *People with Down's syndrome inherit a whole extra chromosome. They have three copies of chromosome number 21, instead of the usual two. This causes severe mental disabilities, but these people are also reknowned for their loving natures and good humour.*

Some diseases, while still not easy to treat or cure are much easier to understand. One of these is sickle cell anaemia, a fatal blood disorder that commonly affects people who originate from tropical Africa and the Mediterranean countries.

Oxygen is carried around our bodies by haemoglobin, a protein in our red blood cells. In sickle cell anaemia, where the base sequence in the haemoglobin gene should read GAG, it reads GTG instead. That single base change causes the production of an incorrect amino acid – valine instead of glutamic acid – which is built into the haemoglobin molecule. The resulting malformed haemoglobin dissolves less easily than normal haemoglobin. It forms crystals in the red blood cells, deforming them and giving them a characteristic curved or sickled shape. These sickled cells clump together and clog the arteries.

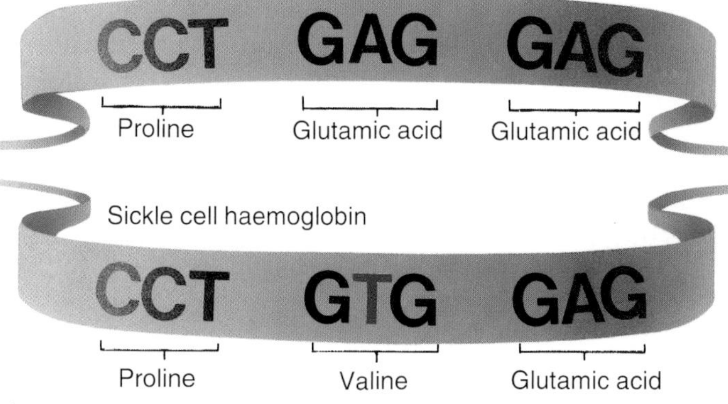

△ *Sickle cell anaemia starts with a mistake in just one letter amongst the thousands on the gene which codes for haemoglobin. As a result, the amino acid Valine is substituted for Glutamic acid. Inheriting two copies of this faulty gene causes the disease.*

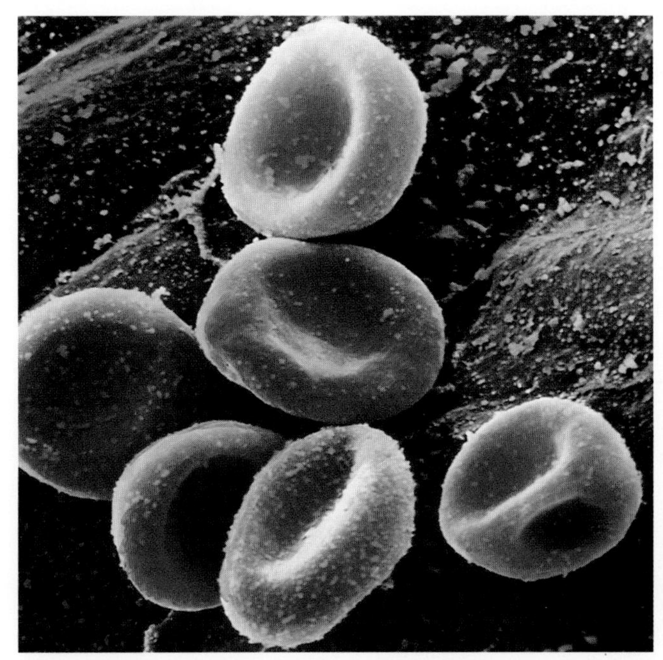

△ *Red blood cells that contain normal haemoglobin have a characteristic rounded and dimpled shape.*

△ *These cells have been curved into the shape of a sickle by the crystals of malformed haemoglobin that they carry.*

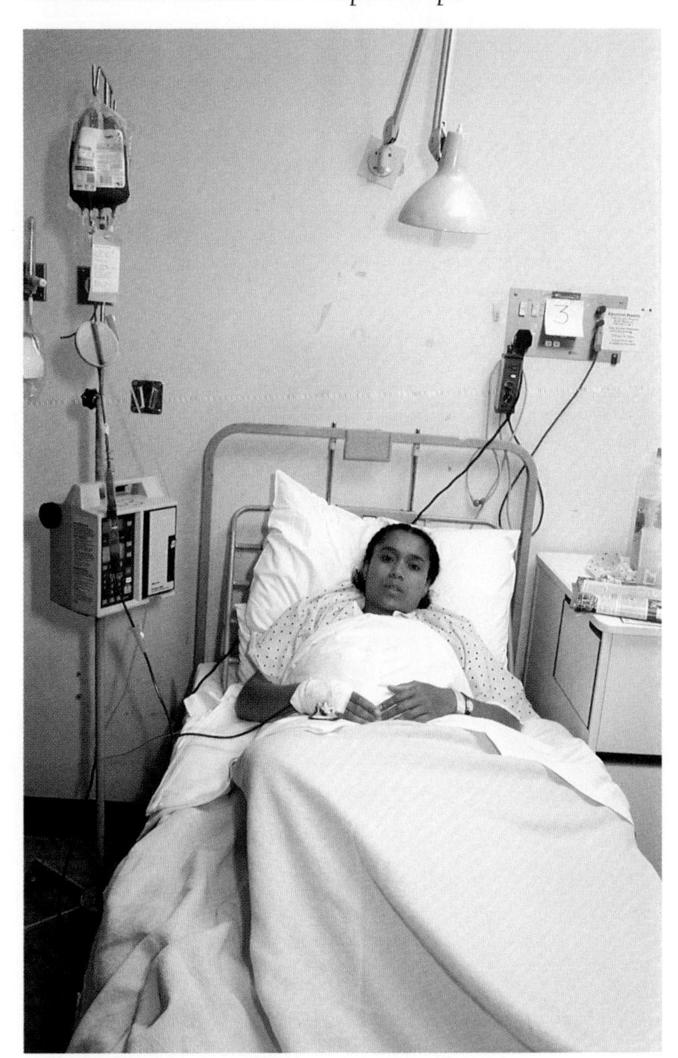

△ *Sickle cell disease causes chronic anaemia, and sufferers may need frequent blood transfusions.*

Meanwhile, the sufferer's body attempts to destroy the deformed sickled blood cells, and this causes serious anaemia. People with sickle cell disease do not usually survive beyond the age of twenty.

Although inheriting two copies of this recessive faulty gene causes sickle cell disease, inheriting just one copy is not a problem. In fact, it gives the gene carrier some protection against malaria so sickle cell carriers who live in tropical regions are likely to survive and have children, passing the disease on to the future generations. People who do not carry the gene are more likely to catch malaria, so natural selection has caused sickle cell anaemia to survive in parts of the world where malaria is rife.

Many other genetic diseases are more common in certain parts of the world than others (see pp. 30-31). Sometimes this is due to the so-called 'founder' effect, where colonists who established a new community took with them some genetic abnormality. For instance, the Dutch pioneers to South Africa took with them a condition known as Van Beeker's syndrome. In other cases, there is no obvious reason why a genetic disease is more common in some areas than others. This is true of cystic fibrosis, which is much more prevalent in northern Europe than in other parts of the world.

There are only a few diseases where, as with sickle cell anaemia, the abnormal gene responsible has been identified and the precise 'error' has been pinpointed. But genetic research is progressing rapidly, and every year sees more inherited diseases traced back to their genetic cause.

A WORLD VIEW

INHERITED diseases crop up with different frequencies in different populations around the world. Sometimes this is because in certain conditions a defective gene can give an advantage to someone who carries just a single copy. Other groups inherited their problems from the few people that founded their population many years before – the founder effect. However, now that people and their genes move freely around the world, local pools of problem genes are being diluted.

Across the USA, diseases are associated with ethnic groups rather than with geographic location. North Europeans brought CF with them, Africans brought sickle cell anaemia and Ashkenasi Jews brought Tay Sachs disease. However, the incidence of Tay Sachs amongst one group living in Brooklyn has been cut by 75 per cent by having the community matchmaker avoid pairing known carriers of the disease.

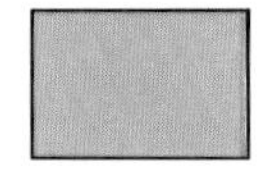
CYSTIC FIBROSIS

MIXED RACE

SICKLE CELL

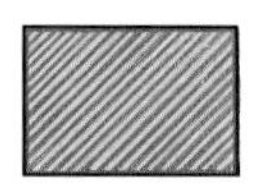
TAY SACHS

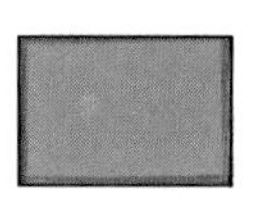
THALASSAEMIA

In the 19th century, 100 times as many deaf people were born in Martha's Vineyard as in the rest of America. Many people there were descended from three British colonists who arrived in 1634 and who probably all carried a gene for deafness. Breeding within the community resulted in many people inheriting two copies of the gene and the deafness that goes with that. (Since then the gene pool has been diluted and the incidence of deafness has gone down).

When the Dutch settled in South Africa they took with them the gene for Van Beeker's syndrome, a bone disease which kills in early adulthood. This condition is hardly known outside Afrikaner groups in South Africa and parts of Holland.

In northern Europe, 1 in 25 people carry the gene for cystic fibrosis, and 1 child in every 2,000 live births suffers from the disease. CF may have become so common because a single copy of the gene protects against diahorrea, but this does not explain why the gene's incidence elsewhere is so low. In oriental populations only 1 in 100,000 live births is affected.

One in 2,500 Ashkenasi Jews of East European origin are affected by the devastating Tay Sachs disease wherever they live – 100 times higher incidence than in any other population. Sufferers rarely survive beyond the age of three. This is another example of the founder effect, and the incidence is reducing with time.

About 1 in 50 live births in parts of Africa are affected by thalassaemia, a lethal form of anaemia. As with sickle cell disease, carriers with one copy of the gene are protected from malaria. The disease used to be common all around the Mediterranean, but screening and counselling have almost wiped it out.

Sickle cell anaemia affects 1 in 50 live births in tropical Africa. Inheriting just one copy of the sickle cell gene protects the carrier from malaria, so sickle cell carriers have a vital advantage where malaria is common.

SEARCHING FOR THE GENE

In most cases, we still do not know which genes cause which diseases, and which predispose people to develop a problem. Some of the most important work today for geneticists is in tracking down the problem genes in our DNA, discovering which proteins the genes are responsible for, and then finding out exactly what has gone wrong with the sequence of the base pairs.

The discovery of the gene that is responsible for a particular disorder often makes it possible to develop a test to identify people who carry the disease. Finding out which protein that particular gene manufactures significantly increases the chances of understanding the nature of a disease and of finding ways of treating it, either by conventional means or by gene therapy (see pp. 40-41).

Throughout our genes there are particular stretches of DNA which are easily recognisable, and the exact location of which is known. They serve no apparent biological purpose, but they can be used by geneticists as signposts, or markers, to guide them as they hunt around the genetic material. One very common type of genetic marker, part of our 'junk' DNA, is a short sequence of bases repeated a different number of times in different people. These are the same markers as are used in DNA fingerprinting (see pp.26-27).

When searching for the gene responsible for a specific inherited disorder, geneticists often begin with families in which many members have the disease. They analyse DNA from as many of the family as possible. If they are lucky, they will find that those with the disease carry a particular marker stretch of DNA, and those without it do not. Where this is the case, it is probable that the defective gene is not only on the same chromosome as the marker, but also very close to it.

▽ *This scientist is examining cells collected from families whose members inherit a number of genetic diseases. The aim is to determine which gene causes which disease.*

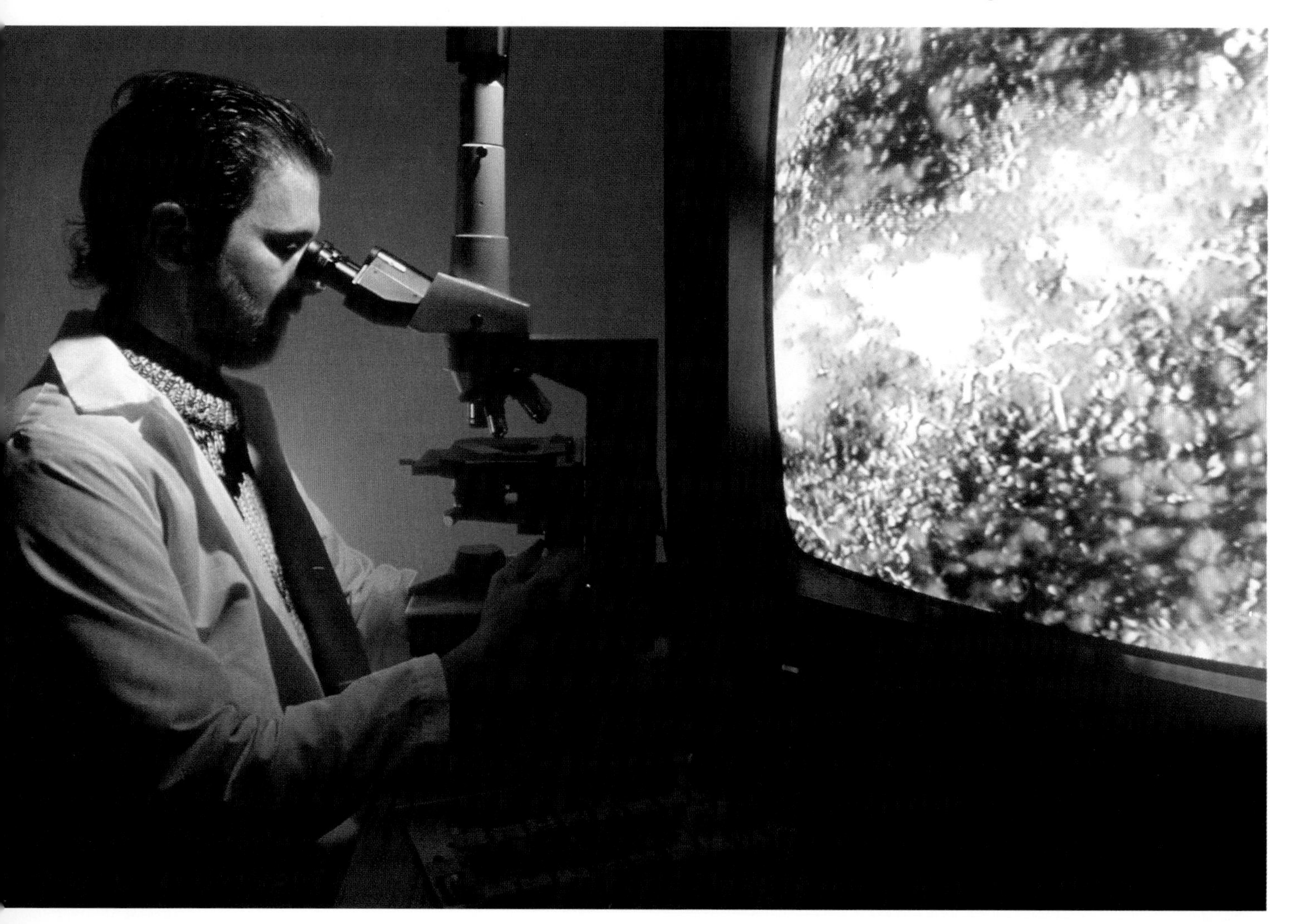

The genes within each pair of chromosomes get mixed up whenever an egg or a sperm is formed. Stretches of DNA that were once near one another on the same chromosome may end up on different chromosomes and would then no longer be inherited together. The closer together a gene and a marker are, the less likely this is to happen. If they are so close together that they are virtually never separated, they are said to be linked. Discovering a link between an identifiable marker and a disease guides geneticists to the chromosome that carries the disease gene, and to its approximate position on the chromosome. Such a link also makes it possible to develop a test to establish if a defective gene has been inherited.

To isolate genetic markers, DNA molecules are cut by special bacterial enzymes that always snip at the same spot in a base sequence. For instance, the enzyme in this example always cuts the DNA between T and C in the sequence GAGCTC. This cuts two marker fragments from the DNA, one from each chromosome in a pair. In the mother's case, this produces one long fragment and one short one – the exact length depends on the number of repeats within each fragment. Treating the father's DNA would also result in two fragments, but these would be different lengths from the mother's. Any child would inherit one of these markers from each parent.

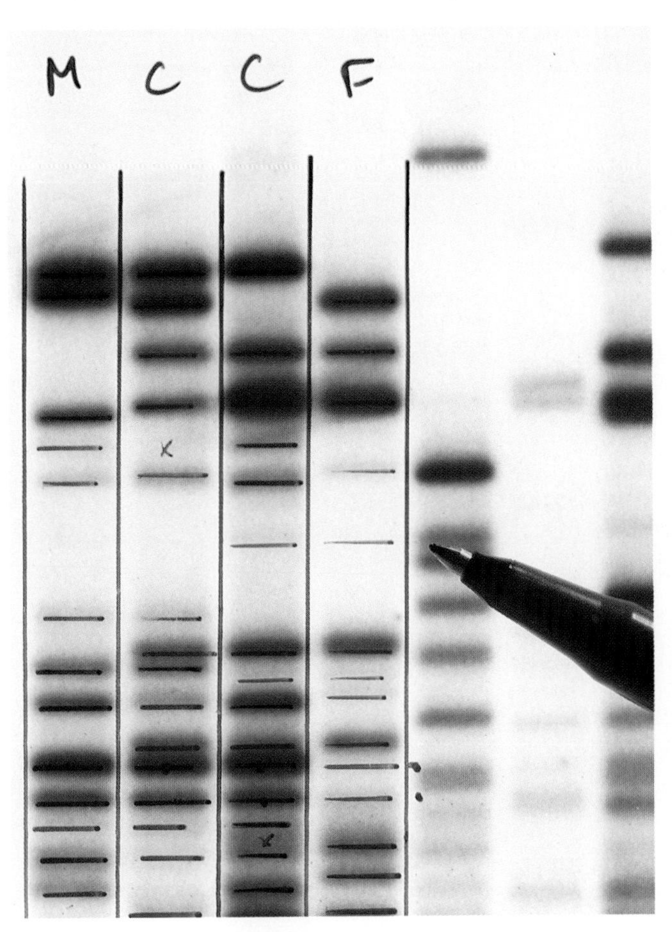

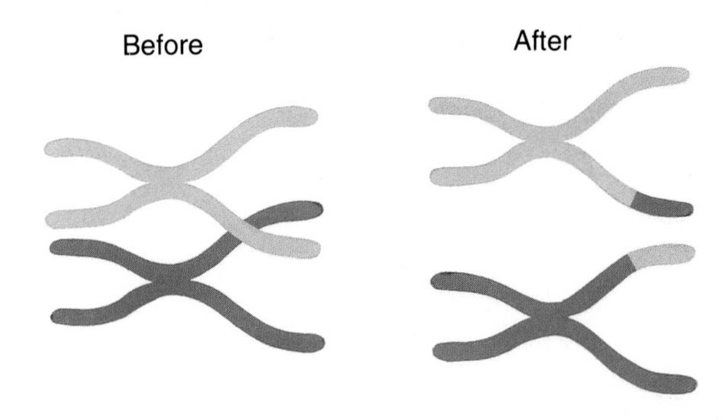

△ *The closer together a gene and a marker on a chromosome, the less likely they are to be separated when new eggs and sperm are formed.*

CHROMOSOME PAIR (FROM MOTHER)

GAGCTC | TGAT | TGAT | TGAT | GAGCTC

Chromosome 1

GAGCTC | TGAT | GAGCTC

Equivalent region on Chromosome 2

MARKER FRAGMENTS CUT BY ENZYME

Enzyme Enzyme

GAGCT C | TGAT | TGAT | TGAT | GAGCT CGG

Chromosome 1 gives long marker fragment

Chromosome 2 gives short marker fragment

△ *The repeating sequence TGAT is a well-known marker. Cutting the sequence at each end of the repeating section produces a fragment whose length depends on the number of repeats.*

These strips from a mother, father and two children show how the DNA fragments of different length have been inherited. The shorter and lighter the fragment, the further along the strip it will be carried by the test procedure. If the mother carried a defective gene that was linked, say, with a shorter fragment, then the fact that child 1 has inherited the fragment would mean that the child has also inherited the problem gene. Finding a marker that is linked to a defective gene is usually the first step on the way to finding the disease gene itself.

◁ *These are DNA strips from the four members of a family. Every band on the children's DNA strips (C) matches one on either the mother's (M) or the father's (F) strips. Each band corresponds to a particular stretch of DNA called a marker.*

TRACKING DOWN DISEASES

THESE PHOTOGRAPHS were taken by a North American research team when they set out find the gene that causes the inherited disorder Huntingdon's Chorea.

Venezuela 1981. Dinnertime in a small, hot, crowded house in a water village on Lake Maracaibo. A group of American researchers were trying to persuade a reluctant fisherman to give them a blood sample. The rest of his family were there eating, listening to music and teasing him – they had all give a sample and so should he.

The researchers were part of a team searching for the single dominant gene that causes Huntingdon's Chorea, a dreadful condition where the whole nervous system degenerates, leading to uncontrolled, jerky movements, dementia and ultimately, 10 to 20 years later, death. When the researchers started their work, they had no idea where in a sufferer's genetic material they should look for the responsible gene. They hoped to find a known genetic marker (see pp.32-33) which was being inherited with the disease and which would guide them to the gene for Huntingdon's Chorea.

△ In the water villages of Lake Maracaibo, about 100 people have Huntingdon's Chorea. They are all members of the same family.

▽ People with Huntingdon's Chorea have no obvious chemical abnormalities in their bodies. But as the condition develops their personality changes and they suffer characteristic, involuntary twitches.

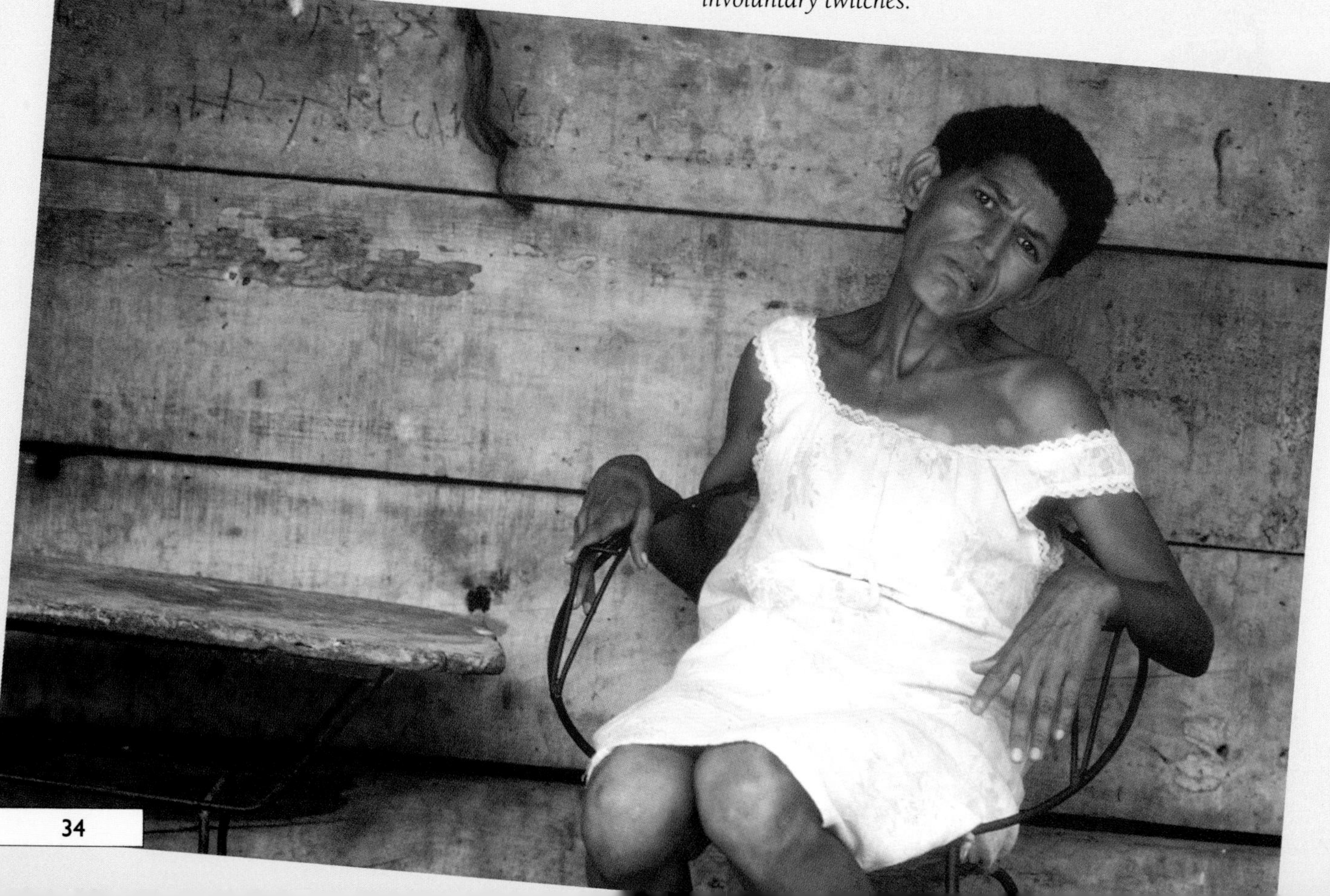

Huntingdon's Chorea is caused by a dominant gene. Any person with an affected parent has a 50 per cent chance of inheriting the disease themselves. But in the normal course of events, symptoms do not develop until sufferers are somewhere between the ages of 35 and 70. Until then, they have no idea whether or not they will develop the disease. By that time they may well already have had children and may or may not have passed on the Huntingdon's gene.

When the quest for the Huntingdon's gene started, no one had any idea what chromosome it might be on, let alone its exact base sequence. While analysing blood from families which carried the disease, the researchers put the marker theory into practice. The larger the affected family, the more reliable any marker they found was likely to be.

The disease is found all over the world, but amongst the "family" of 2,500 people on Lake Maracaibo, the defective gene is particularly common. About 100 of them are already afflicted with Huntingdon's Chorea, and another 1,100 young people and children have a 50 per cent chance of developing it. The research team gathered 570 blood and skin samples from which the genetic material was extracted and analysed. They noted on a family tree which people had the disease, and which people carried various genetic markers.

△ *The villagers were interviewed to determine how they fitted into the family tree.*

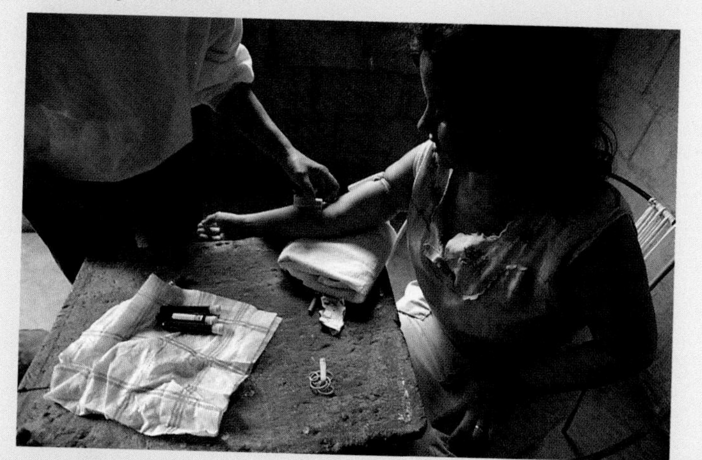

△ *Nancy Wexler, who instigated the search for the Huntingdon's gene, with the enormous family tree that she helped draw up.*

In March 1983, the team identified a marker so tightly linked to the Huntingdon's gene in all affected families that any family member carrying it has a 1,600 to one chance of carrying the gene for the disorder as well. This allowed the development of a test that can predict with some certainty whether or not a person will develop Huntingdon's Chorea. This was the first time that such a marker had been found for any genetic disease, and the first time such a predictive test was possible.

But it took ten more years to identify the gene responsible. People with Huntingdon's Chorea have no other apparent abnormality, and the gene in which the defect has been found was not known before. Now researchers are working to find out what the protein it produces usually does, and why a mistake in it causes the disease.

The defect itself is the lengthening of a repeating sequence of three base pairs – CAG, the triplet (see genetic code pp.11) that specifies glutamine – at one end of the gene. The length of this repeat seems to correlate with the onset of the disease; the more repeats there are, the earlier the disease is likely to strike. Soon, it may be possible to discover not only who will develop Huntingdon's Chorea, but when that might happen as well.

However, many people do not want to take the test. While a negative result brings great relief, if the test is positive there is nothing that can be done to stop the disease developing. The only advantage of early knowledge is that a person can take it into account when deciding whether or not to have children. They can also receive counselling to help them cope with the difficult years ahead, but as yet nothing can help them avoid the disease.

◁ *The researchers took thousands of blood samples. Some of the villagers had to be persuaded, but other became angry when they thought they might be left out.*

A FAMILY AFFAIR

Dominant inheritance

Huntingdon's Chorea is caused by a dominant gene.
One parent carries a single defective gene which dominates its normal counterpart. Each child has a one in two, or a fifty-fifty chance, of inheriting the gene, and statistically half the children will be affected.

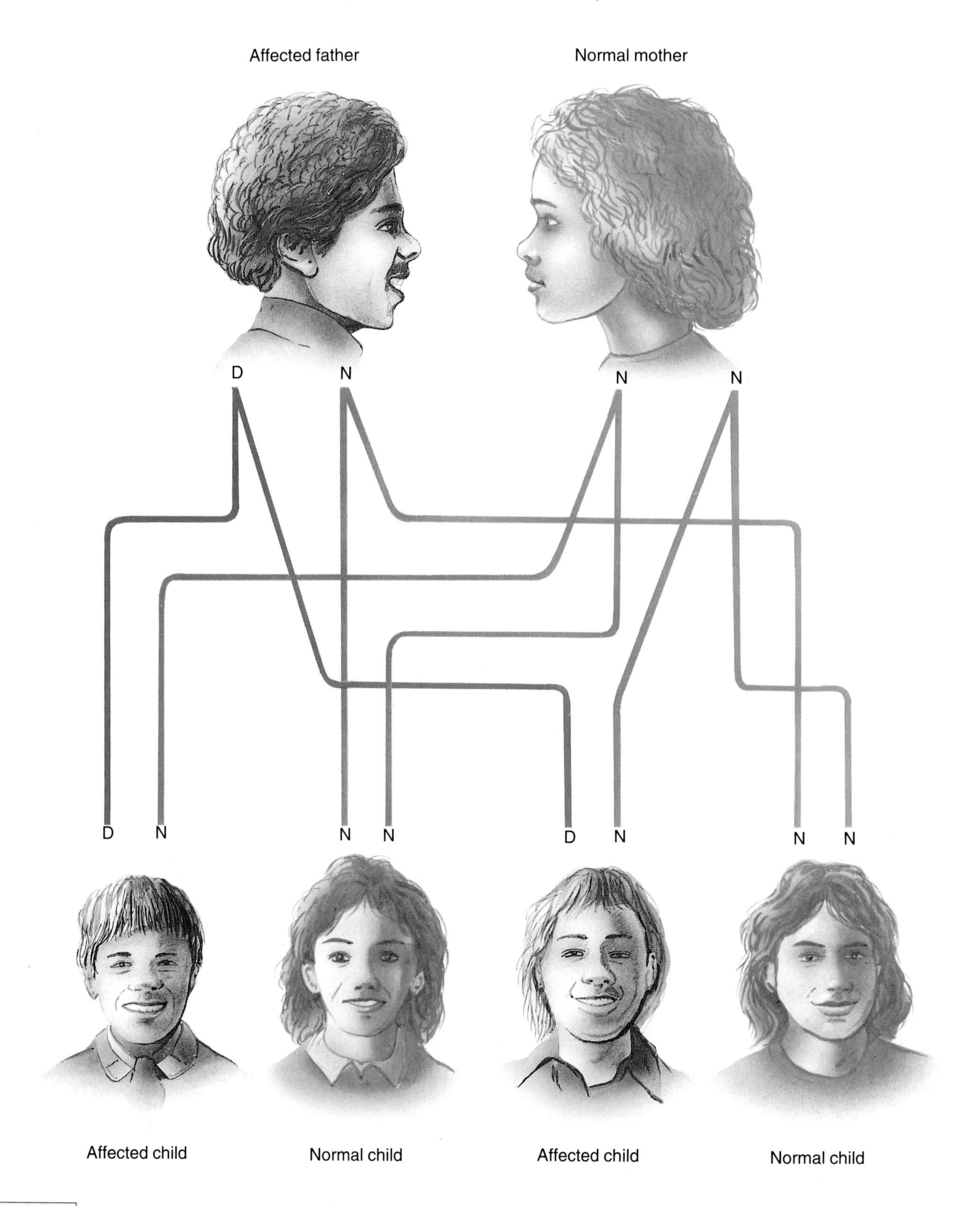

Recessive inheritance

Cystic fibrosis is caused by a recessive gene which in Northern Europe is carried by one person in every 25.

Where both parents are carriers of the disease they each have one normal copy of the gene and one defective copy. In these cases, the normal gene is dominant, so the single healthy gene is enough for normal function, and neither parent has any symptoms of the disease. The disease will only develop where *both* copies of the gene are defective. If both parents are carriers; their children will have a 1 in 4 chance of inheriting two faulty genes and thus developing the disease. Statistically, one quarter of all their children will have the disease, one quarter will carry two normal genes, and half will be carriers.

GENETIC SCREENING

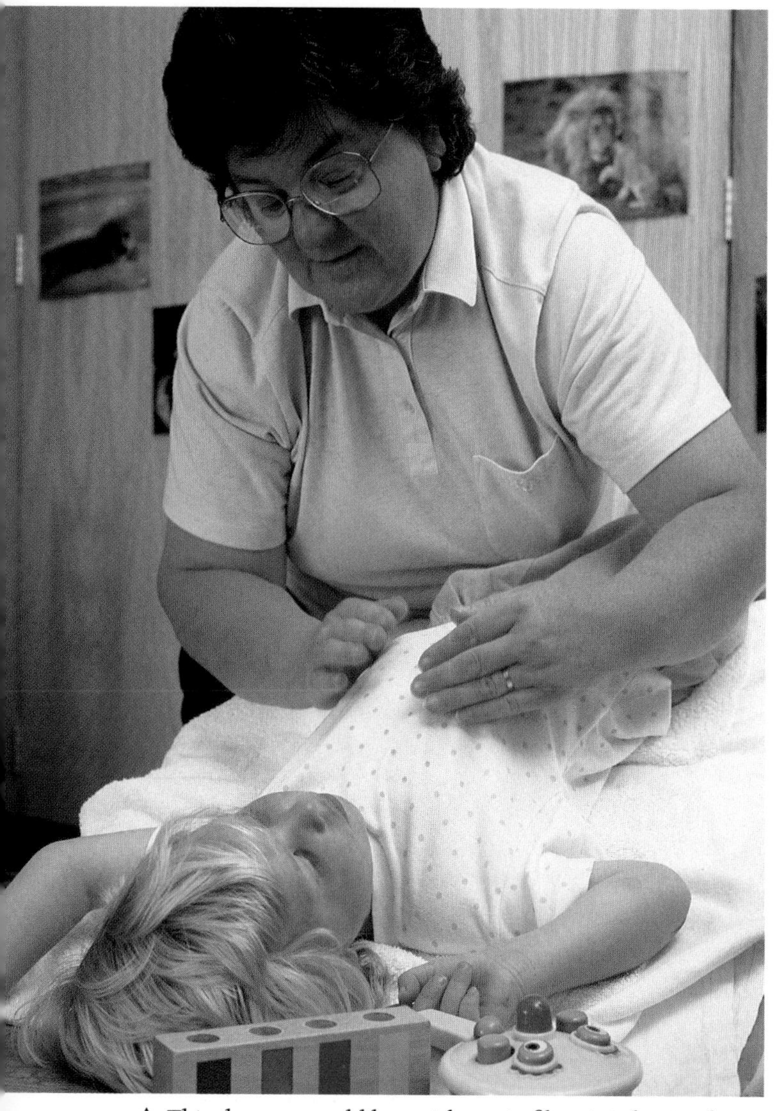

△ *This three-year-old boy with cystic fibrosis is having his daily dose of physiotherapy.*

AS MARKERS and defective genes are discovered, it is possible to search for an increasing number of diseases and predispositions, for instance to certain types of cancer. The test for Huntingdon's Chorea made possible by the discovery of the genetic marker was the first time doctors could look into a person's genes and predict the future.

Predictive tests throw up some very difficult questions. In the case of Huntingdon's Chorea, for instance, there is no cure or even treatment available, and it seems clear that a potential carrier has the right to choose not to take a test. But what if they already have children and one of them wants to know if they carry the disease? If the child takes the test and the result is positive, that confirms not only that they will one day develop Huntingdon's Chorea, but that the parent from whom they inherited the disease is certain to as well. So which right should prevail? The right to know or the right not to know?

The availability of genetic tests for a range of diseases will lead to many other practical and ethical dilemmas (see pages 44-45). For instance screening to find carriers of genes for recessive disorders brings with it a specific set of problems.

In Britain, five children are born every week with cystic fibrosis. All their lives, they must have regular, strenuous physiotherapy to shift the sticky fluid that forms in their lungs, and few who have the disease survive beyond their late twenties. To contract the disease, a child must inherit two copies of the defective gene, one from each of their parents – parents who may themselves be perfectly healthy.

Programmes to screen for the 1 in 25 people in the general population who carry the cystic fibrosis gene, are now being trialled in Britain and the United States.

The test itself is very simple, and the genes are extracted from a mouthwash sample. But the potential implications are rather more disturbing. Before taking the test, everyone is counselled about what being a carrier means. Couples who are both found to be carriers, and who are thus highly likely to have an affected child are offered extra counselling to help them consider all the options.

◁ *Giving a sample to screen for carriers of cystic fibrosis is as simple as spitting into a bottle.*

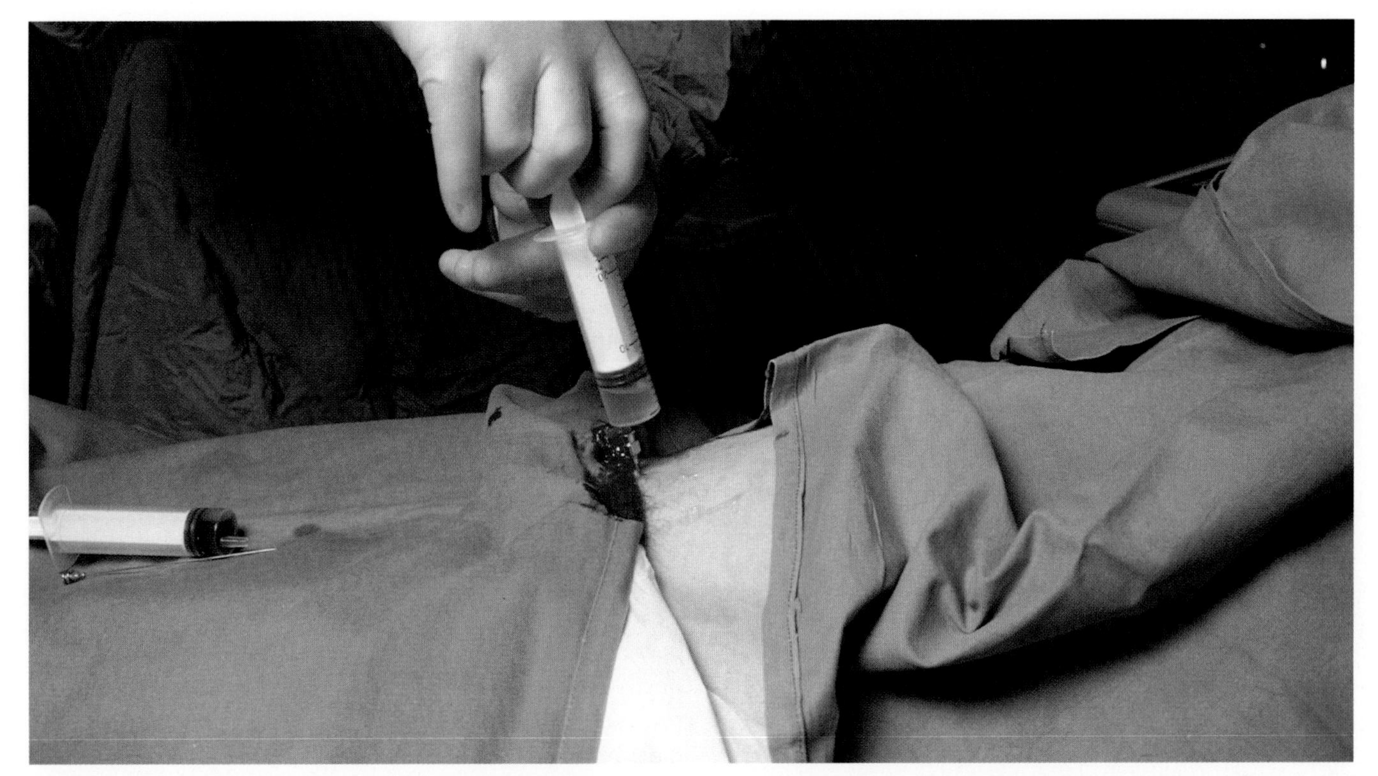

△ *The genes of a foetus can be screened by analysing a sample of amniotic fluid drawn through a needle from the mother's womb.*

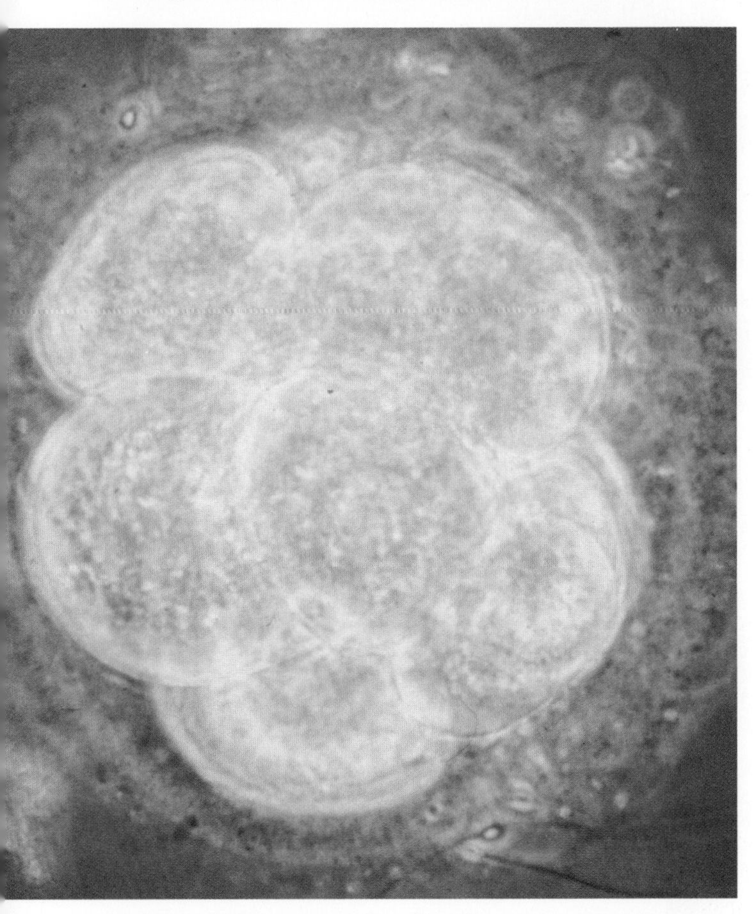

△ *If one cell from this eight-cell embryo were taken away for screening and found to be unaffected, the remaining seven could still be inplanted into the mother's womb to develop into a normal baby.*

A couple who are both carriers have to consider whether or not they want to take the 1 in 4 risk of having a child with cystic fibrosis? Will they go ahead with any pregnancy and care for a child born with the disease? Or will they have the foetus tested and consider abortion if that test shows that the baby will be affected? They might decide not to have children of their own, and turn instead to adoption.

Other screening techniques are being developed for a range of diseases that take some of the agony away from deciding the future of an unborn child. Doctors at the Hammersmith Hospital in London have pioneered a technique to screen 8-cell embryos. After test tube fertilisation, one or two cells are removed from the embryos and screened for defective genes. Then only embryos with no apparent defects are implanted in the women to develop into normal, healthy babies.

Another new approach goes back even further along the reproductive chain. Researchers in America are screening the unfertilised eggs of women suffering from a severe enzyme deficiency to exclude eggs that contain faulty DNA. Around the world, groups are trying to apply the same technique to screening for sickle cell anaemia, cystic fibrosis, muscular dystrophy and other inherited diseases. So it could soon be possible not only to screen for these diseases, but to do so without touching a single embryo or foetus.

A THERAPY FOR THE FUTURE

IN THE 1970s, a sterile bubble was used to protect David, who suffered from a severe, incurable disorder of his immune system. The disease left him vulnerable to any passing infection, and in 1984 it finally claimed his life. Just six years later, at 12.52 pm on 14 September 1990, in the USA, the world's first gene therapy was begun on a four-year-old girl with a similar inherited disease, known as ADA deficiency.

Now that so much is known about the genetic basis of disease, we can look forward to using gene therapy to treat, and even cure, some genetic defects. The simplest way to do this is to replace the bad genes with good ones. Once in place, the new genes should take over, producing the protein which the abnormal gene should have been making. In a sense, gene therapy is just a way of moving the point of drug production from a factory to a cell inside the body.

◁ *David had no natural protection from disease. In an attempt to keep him away from infection, he spent most of his life completely enclosed in a sterile bubble.*

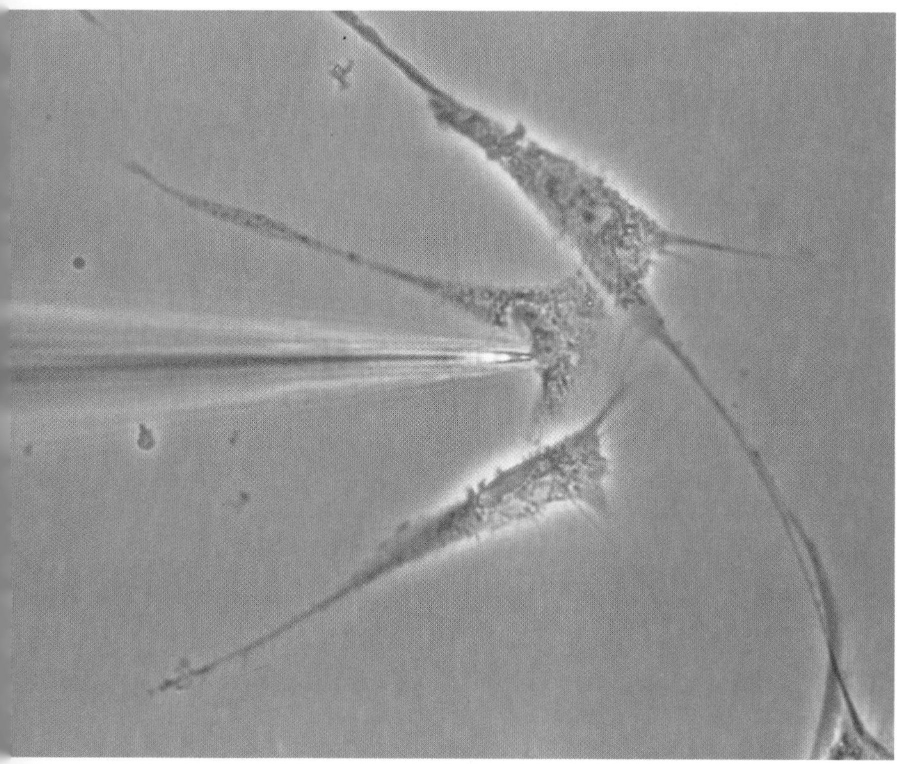

One way of getting a corrective gene into cells that need it is to inject the genetic material directly into the cell. This is an effective, but not very practical process. The most efficient way of delivering genes is to exploit the natural talents of viruses. Viruses are little more than tiny packets of genes wrapped up in protein coats. Normally these genes can force the cells they invade to create new viruses, which then spread and infect other cells. Certain types of virus merge their own genes into the cell's genetic material.

◁ *Corrective genes can be injected directly into human cells using a micro-pipette that is worked by remote control. These T cells are being injected with the healthy ADA gene.*

To treat the little girl, with ADA deficiency, the gene doctors use just such a virus. First they make the virus safe by destroying the genetic material it needs to reproduce itself. They then engineer the ADA gene into this modified virus. When the virus infects the target cells, in this case T cells from her immune system, it transfers the healthy ADA gene into the girl's DNA.

The science is complicated, but the treatment is straightforward. Once a month, the girl receives an injection of a greyish liquid into the back of her hand. This liquid carries about one billion T cells, each containing a copy of the normal gene that her body is lacking. After three years of therapy, these cells were producing enough enzyme to support the girl's devastated immune system.

By exploiting viruses that attack different parts of the body, this technique is being extended to treat many different conditions. But it is not the only way to make gene therapy work. By 1993, over 50 gene therapy trials, using different methods to tackle different disorders, had been approved in the United States alone. These include therapies for liver disease, AIDS and many cancers.

△ *In ADA therapy, healthy ADA genes are put into T cells taken from the patient's blood. Because these cells only survive for a month or so, the treatment has to be repeated regularly.*

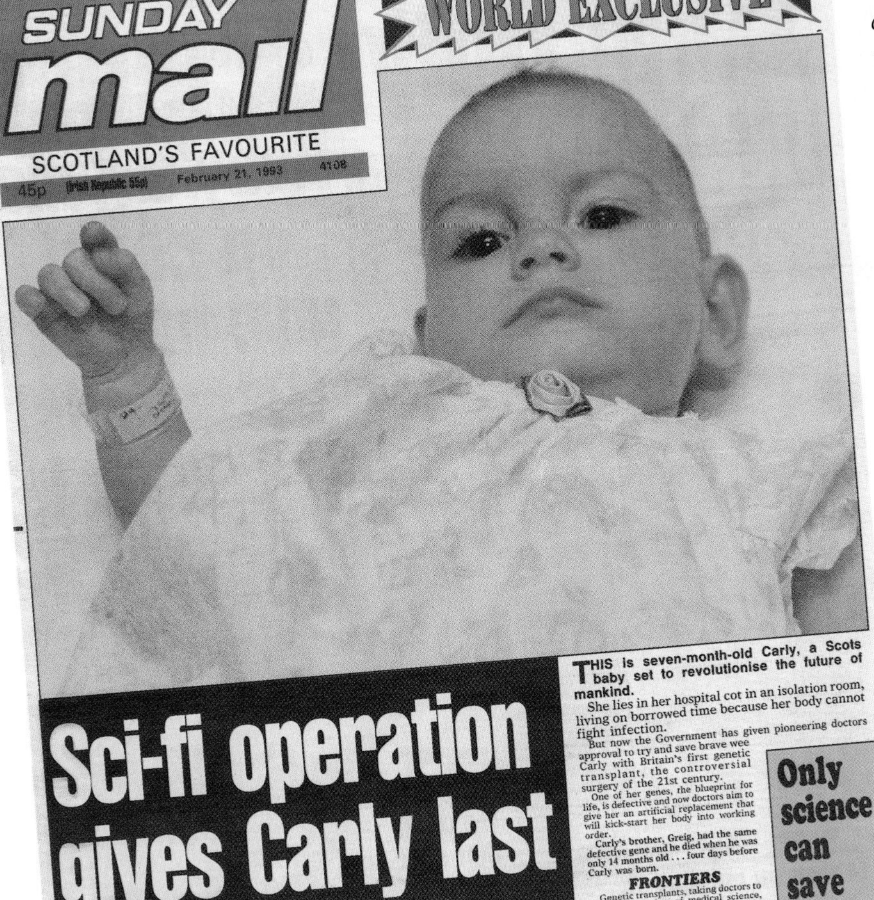

SUNDAY mail

SCOTLAND'S FAVOURITE

45p (Irish Republic 55p) February 21, 1993 4108

WORLD EXCLUSIVE

Sci-fi operation gives Carly last chance of life...

THIS is seven-month-old Carly, a Scots baby set to revolutionise the future of mankind.

She lies in her hospital cot in an isolation room, living on borrowed time because her body cannot fight infection.

But now the Government has given pioneering doctors approval to try and save brave wee Carly with Britain's first genetic transplant, the controversial surgery of the 21st century.

One of her genes, the blueprint for life, is defective and now doctors aim to give her an artificial replacement that will kick-start her body into working order.

Carly's brother, Greig, had the same defective gene and he died when he was only 14 months old . . . four days before Carly was born.

FRONTIERS

Genetic transplants, taking doctors to the very frontiers of medical science, could make certain cancers, multiple sclerosis, cystic fibrosis, muscular dystrophy and other conditions triggered by faulty genes obsolete in future generations.

But it is the kind of medicine which also raises doubts over population control and people made to order.

Only science can save her

PAGES 2, 3 and 4

For some people genetic treatments may be turning into cures. When she was just 9 months old, Carly Todd, who also suffers from ADA deficiency, was too small to withstand the regular monthly injections of white blood cells that she needed. Instead, doctors in London tried to insert the healthy gene into very young cells in her blood from which all other blood cells stem. These stem cells are very rare and very hard to find. If the doctors succeed in infecting her stem cells, Carly will never need another treatment for ADA. Even if they do not manage a cure, the doctors hope to alter cells that will last at least a few years.

◁ *In February 1993, little Carly Todd became the first person in Britain to receive gene therapy.*

TESTING, TESTING...

SINCE THE first gene therapy experiment began in 1990, research in the field has exploded. Now therapies are being devised for diseases as diverse as Alzheimer's, AIDS and arthritis.

In April 1993, geneticists in the United States first used the common cold virus in a gene therapy for people with cystic fibrosis (CF). Modified so that it should not cause chest infections, the virus delivers corrective genes into the lungs of people suffering from CF. The cold virus was chosen because it naturally infects the cells that line the lung. If the treatment works, it will have to be repeated every two or three months as the old lung cells die off and new, untreated ones take their place.

As well as inherited disorders, gene therapy offers hope for incurable diseases like AIDS and, most particularly, cancer. While cancer is not really a genetic illness, it is associated with the expression of faulty genes, and the idea of using gene therapy to combat many different cancers has really taken hold.

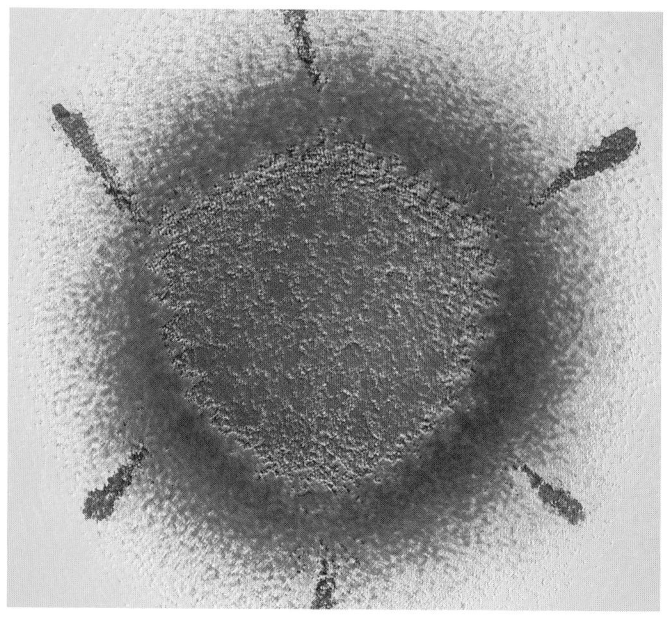

△ *An electron micrograph of the cold virus now being used in trials of a gene therapy for cystic fibrosis.*

△ *Smoking significantly increases your chances of getting lung cancer, especially if you are genetically susceptible.*

Mass gene therapy may well begin with people who have lung cancer. While there is no doubt that smoking triggers lung cancer, the people who develop it are genetically susceptible.

One genetic mutation commonly associated with lung cancer causes the production of a protein that should not be there. This protein promotes uncontrolled cell growth, which results in a tumour. Researchers are making a synthetic gene which interferes with the production of this protein.

▷ *In this false colour image it is clear where smooth, malignant tumour cells (green) have invaded the normal lung tissue.*

A different style of genetic treatment, being tested on people with brain tumours, makes cancer cells vulnerable to attack from an anti-viral drug usually used to treat herpes. In the treatment, a helper virus is used to insert a gene from the herpes virus into the tumour cells. The gene codes for a herpes enzyme which makes infected cells sensitive to the drug. Once in the brain tumour, the helper viruses spread the enzyme. When the patients take the anti-viral drug, it kills cells containing the enzyme and so it should destroy the tumour. So far, the results look encouraging, but it is still early days.

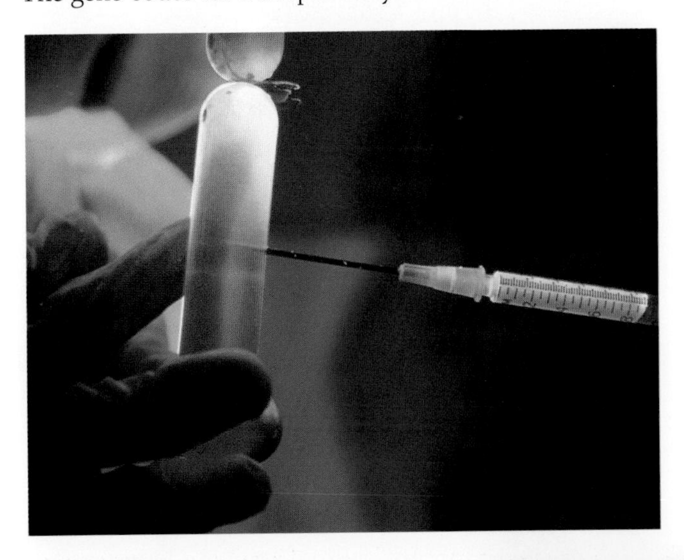

Unfortunately, treatments for some inherited diseases such as sickle cell anaemia, which are known to be caused by a single defective gene, are not ready for trial. Transferring the corrective gene has posed no particular problems, but in animal experiments the transplanted genes have proved very difficult to control. On the other hand, AIDS, which is not inherited, is proving to be an approachable target. This is because gene therapy can be used to produce any protein, even ones our bodies do not produce naturally. By selecting the right protein, it may be possible to boost the action of the immune system.

◁ *Genetic therapies are still in the very earliest stages of development. The final treatments will be very different from those being tested today.*

If gene therapy were applied to the cells that make eggs and sperm, it could remove the threat of genetic disease from future generations. However, if a serious error occured during treatment, it would be handed down to all the patient's descendants. Also, our understanding of disease genes and why some are common is still too limited for anyone to be sure that removing them is a good idea. If a couple who carry an inherited problem want to guarantee the genetic health of their children, it is easier and safer to screen their fertilised eggs and then to reimplant only the unaffected ones.

Genetic alteration of eggs and sperm is more likely to be considered when we can engineer resistance to infectious diseases like AIDS. However, this is not yet possible, and for the time being so-called germ line gene therapy, is generally agreed to raise too many ethical problems to be acceptable.

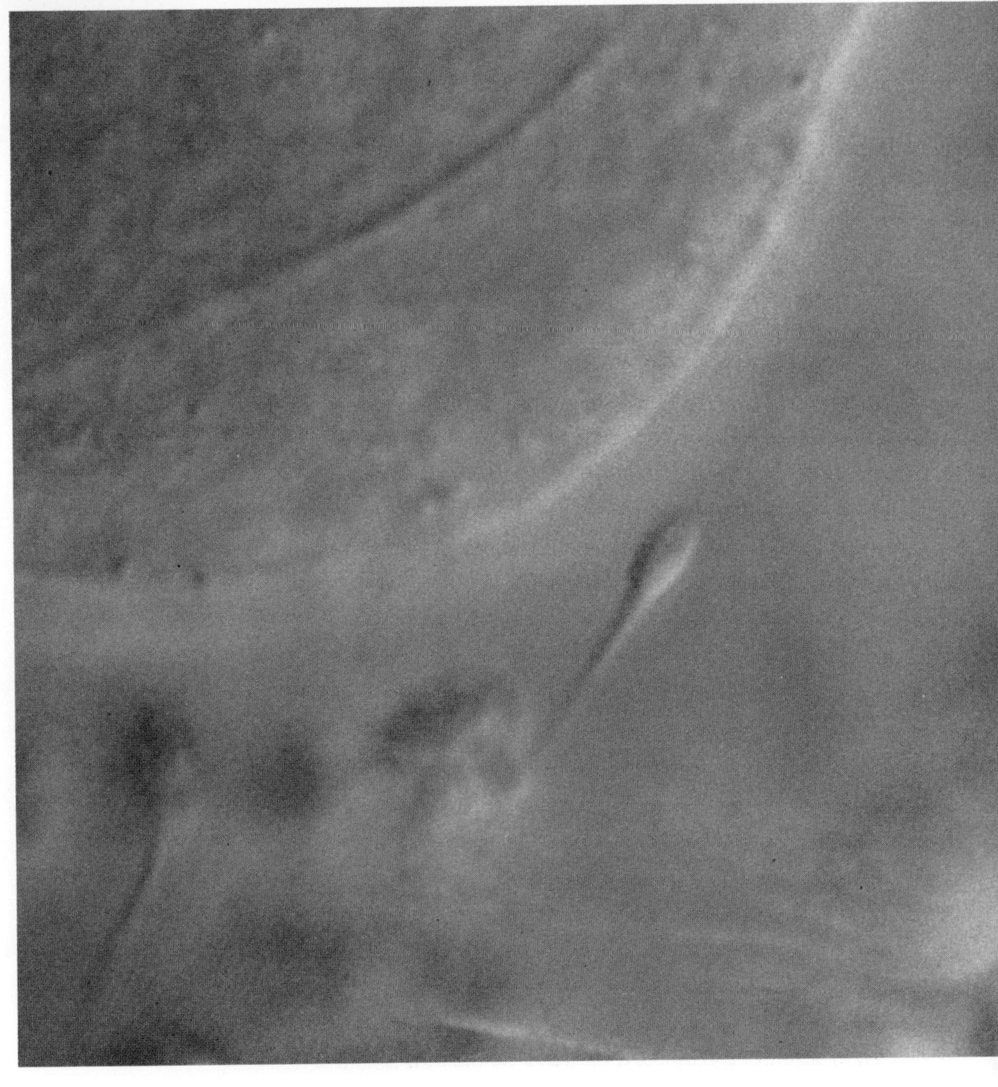

▷ *Changes made to the genes of eggs and sperm would affect all future generations. Although technically possible, such a alteration of our genetic code is not yet ethically acceptable.*

LOOKING AHEAD

IT WAS only in 1953 that Francis Crick and James Watson worked out the structure of the DNA molecule – the now famous double helix. Since then genetic research has moved faster than anyone could have imagined.

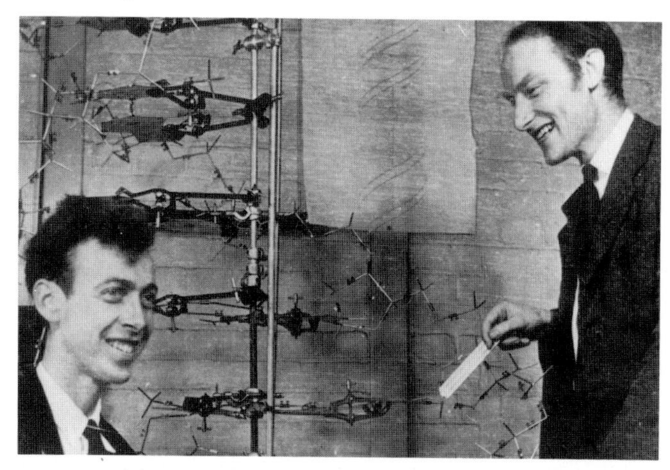

△ *Watson (left) and Crick (right) discovered the physical structure that is the basis of life.*

Recent advances prompted the setting up of the most important research project biology has ever seen – the search for who we are. Geneticists from all over the world are collaborating to find out the exact order of the three thousand million bases in our genetic material, the genetic blueprint we call the human genome. Already almost 10,000 genes have been pinpointed to specific places on our chromosomes. However, geneticists know the function of fewer than 4,000, and with genes themselves making up only 3 per cent of our genetic material, there is still a long way to go.

Knowing the complete sequence of all the human genes and what they all do would not only help to track down genes involved in particular diseases. Geneticists hope it will lead to a proper understanding of how genes function and how they work together to cause predispositions to diseases such as cancer, heart disease and diabetes.

In just a few years, genetic tests able to spot these predispositions as well as inherited diseases, will become available through our local doctors. If we choose to take such tests, who should have access to the results? The results could serve as a useful warning and move us towards a healthier lifestyle, but a positive result could also spoil job prospects, or the chance of buying life insurance.

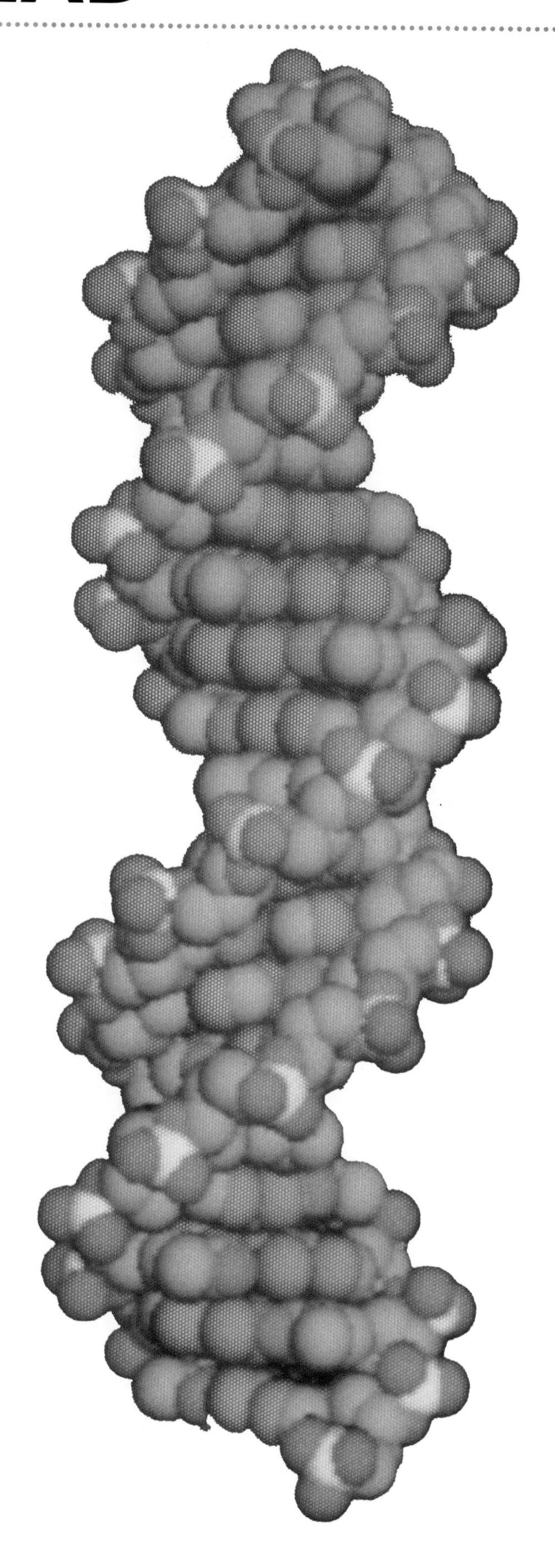

▷ *A computer-generated image of the twisted ladder of DNA.*

And then there are our children. Already we take for granted tests for screening foetuses in the womb and many found to be carrying abnormal genes are aborted. How much further should we go in trying to manipulate the genes of our children?

Geneticists may well discover genes which influence characteristics other than disease – for example, how clever we are, how tall, or even the way we behave. If they do, what should we do with that information?

In a survey conducted in the United States in 1992, the American public expressed a remarkable willingness to use gene therapy not just to treat diseases, but to change other inherited characteristics. The vast majority, 89 per cent, were in favour of gene therapy, 42 per cent also said they would use genetic techniques to improve children's intelligence, and 43 per cent would use it to improve their physical characteristics.

A similar survey conducted in Britain in 1993 produced rather different results. Again the majority, 63 per cent, thought it was right to manipulate the genes of future generations to treat disease. But only 8 per cent thought it acceptable to use genetic techniques to increase intelligence. And a tiny 1 to 2 per cent thought it right to try to improve physical characteristics.

What all this suggests is that the world may soon face a dilemma. The technology of gene therapy, developed with public approval to cure diseases, will almost certainly become a tool which could allow us to modify ourselves in many other ways. But how will we decide what is acceptable and what is not? And will those questions receive the same answers in different parts of the world, and in different cultures.

Cracking the code in which our genetic information is written has been one of the greatest achievements of the twentieth century. Making the best use of that information is going to be one of the hardest tasks of the twenty-first.

▽ *What lies in store for the children of the future, and how much should we be allowed to determine in advance?*

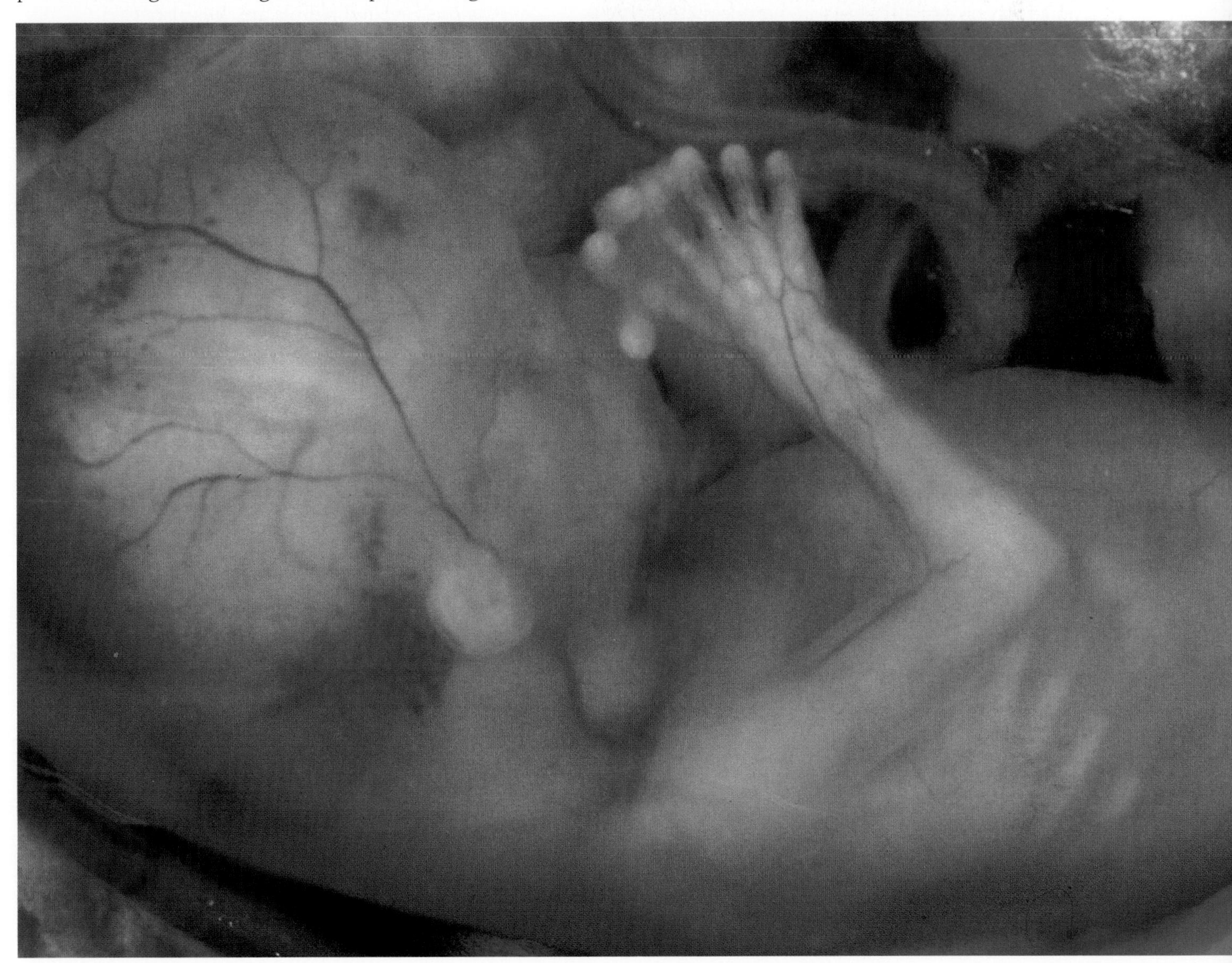

GLOSSARY

antibodies An important part of our immune system. They make inactive foreign bodies such as bacteria and viruses and help to protect us from disease.

bacteria Single-celled living creatures which are too small to be seen without a microscope. Most bacteria are harmless but some cause diseases like tonsilitis, tetanus and pneumonia.

base In genetics, a base is one of four chemical compounds (cytosine, guanine, thymine or adenine) which are part of the DNA molecule and which enable it to carry genetic information.

bioreactor A container in which living things make useful biological materials on a large scale.

cell nucleus The structure at the heart of a cell where chromosomes are found.

chromosomes Microscopic bundles of DNA which carry thousands of genes in sequence along their length. There are normally 23 pairs of chromosomes in a human cell.

DNA Deoxyribonucleic acid. It is the long, thread-like material from which genes are made.

DNA fingerprinting A laboratory technique which compares inherited patterns of genetic markers to reveal how closely related people are. It is also used to identify criminal suspects from their biological samples.

dominant gene A dominant gene forces an organism to exhibit a certain characteristic, even if only one copy of the gene is present. It masks the effect of any alternative form of the gene which may also be present.

embryo The unborn offspring in the very early stages of pregnancy. In humans this means from conception until the eighth week of pregnancy, when the embryo becomes recognisably human.

fertilization The joining of egg and sperm - or the equivalent cells in plants - to make a fertilized egg that can grow into a new creature.

foetus The name for an unborn offspring in the later stages of pregnancy. In humans, this means from eight weeks to birth, which usually comes at around 40 weeks.

gene A length of DNA that carries a section of the genetic code. Most genes are structural genes and code for one protein. Others act as on and off switches.

genetic marker A known sequence of genetic bases which lies close to a gene and which can be used to track the way that gene is inherited in a family. Genetic markers are also used in DNA fingerprinting.

hormones Protein messengers which control many bodily functions such as growth, metabolism and sexual development.

monoclonal antibody An antibody produced in the laboratory which works against only one specific substance or microbe. They are made by copying, or cloning, a single, selected antibody-producing cell.

proteins Large molecules made of long strings of amino acids. Each protein is the product of one structural gene. Human beings are largely made out of proteins.

recessive gene A recessive gene will only force an organism to exhibit a certain characteristic if there are two copies of the gene present, one on each chromosome in a pair.

RNA Ribonucleic acid. A second kind of genetic material, closely related to DNA. It plays an important role in the translation of genetic information from a sequence of bases in DNA to a functional protein.

stem cells Immature cells which give rise to all the different types of cell found in the blood.

structural genes Genes that code for proteins. Other genes act as switches and turn the structural genes on and off as required.

triplet A group of three chemical bases along a chain of DNA. Each triplet is the code for one amino acid.

INDEX